D0754752
DISCARDED
Richmond Public Library

WALTER E. BILLIET

SMALL GAS ENGINES & POWER TRANSMISSION SYSTEMS

A Repair and Maintenance Handbook

A SPECTRUM BOOK

Prentice-Hall, Inc. Englewood Cliffs, New Jersey 07632

Library of Congress Cataloging in Publication Data

Billiet, Walter E.
Small gas engines & power transmission systems.

"A Spectrum Book."
Includes index.

1. Internal combustion engines, Spark ignition–Maintenance and repair. 2. Power transmission. I. Title. II. Title: Small gas engines and power transmission systems.
TJ790.B54 621.43'4'0288 82-3866
AACR2

ISBN 0-13-814327-7

ISBN 0-13-814319-6 {PBK.}

This Spectrum Book is available to businesses and organizations at a special discount when ordered in large quantities. For information, contact Prentice-Hall, Inc., General Publishing Division, Special Sales, Englewood Cliffs, N. J. 07632.

A SPECTRUM BOOK

10 9 8 7 6 5 4 3 2 1

Printed in the United States of America

Editorial/production supervision and page layout by Fred Dahl/Inkwell
Manufacturing buyer: Barbara A. Frick

Prentice-Hall International, Inc., *London*
Prentice-Hall of Australia Pty. Limited, *Sydney*
Prentice-Hall of Canada, Ltd., *Toronto*
Prentice-Hall of India Private Limited, *New Delhi*
Prentice-Hall of Japan, Inc., *Tokyo*
Prentice-Hall of Southeast Asia Pte. Ltd., *Singapore*
Whitehall Books Limited, *Wellington, New Zealand*

CONTENTS

PREFACE

In recent years, power tools and equipment have become commonplace. Each year more people are buying some type of power equipment for work or pleasure. With ever more people moving to the suburbs, many acquire bigger lots—and the equipment they need to take care of it all:

Power mowers
Lawn tractors
Garden tractors
Snow removal equipment

Plus various other types of power equipment to assist in taking care of lawns, gardens, trees, snow, leaves, and the property in general.

The growing amount of leisure time these days has also created a market for other types of equipment powered by small gasoline engines, such as snowmobiles for winter recreational purposes. In some places, they are used for emergency rescue work in the case of heavy snow storms. It has been indicated that snowmobiling is one of the fastest growing sports, after only tennis and skiing.

With the shortage of fuel and the resulting increase in fuel costs, the growth in the sales of mopeds has been phenomenal. When mopeds were first introduced, they were primarily considered to be recreational machines. Today, in many cases, they are being used as a mode for needed transportation.

With few exceptions all this equipment is driven by small single-cylinder gasoline engines. Servicing this equipment has become a big business. While a number of small engine repair shops are in operation, many limit their services to specific makes or kinds of equipment. Some of them may even be too busy when you need service. In addition service costs have risen to the point that many people *have to* put off having the needed work done, which can result in additional expense at a later date.

The purpose of this book is to present, in a simple manner, the procedures that enable you to do much of the necessary maintenance work. In addition, you should be able to make many of the needed repairs when trouble occurs in the engine and in the equipment that the engine powers. From the manner in which the information is presented, you should be able to easily locate the trouble when problems arise.

Understanding in detail how all these machines operate may appear to be a monumental task. But the book presents a simple, yet comprehensive, study of fundamental operations, maintenance, trouble shooting, and repair procedures. Ultimately you should become aware of the fact that these machines are more *alike* in design and operation than they are different. Many of the differences are primarily in component size and structure.

Through proper maintenance you can generally combine economical operation and reliability. By learning how to make simple adjustments and how to recognize and replace worn parts, you should be able to get many years of trouble-free operation from your machine at minimal expense. Also there is always a certain sense of accomplishment achieved when you have successfully completed a repair or maintenance job on a motorized piece of equipment.

I wish to acknowledge the photocopy work done by William Grinnell, which is used throughout the book.

1 INTRODUCTION TO MAINTENANCE AND REPAIR

SMALL ENGINE APPLICATIONS
TYPES OF SERVICE OPERATIONS
COMPONENTS DRIVEN BY THE ENGINE
CONNECTION BETWEEN POWER PLANT AND DRIVEN EQUIPMENT
CHANGING THE DRIVING FORCE
HOW PULLEYS, GEARS, AND SHAFTS ARE SUPPORTED
SERVICE OPERATIONS THE OWNER MAY PERFORM
BECOMING FAMILIAR WITH THE VARIOUS COMPONENTS
HOW THE BOOK IS STRUCTURED
LOCATING TROUBLES

The purpose of the book is to provide information for the person who wants to perform the necessary maintenance and to make the needed repairs on the various machines powered by small gasoline engines. The book is also of value to the person who just wants to learn more about equipment of this type. With the necessary know-how, you can do most of the required maintenance, trouble shooting, and minor repair operations needed to keep such machines operating in a satisfactory manner.

SMALL ENGINE APPLICATIONS

The most common machines powered by small gasoline engines include lawn mowers, lawn and garden tractors, chain saws, mopeds, tillers, snowmobiles, snow blowers, golf carts, and so on. But small gasoline engines have many other different applications. They can be used to drive almost any kind of small machine—either portable, stationary, or wheeled equipment. Some engines are small and light enough to be completely portable, such as those in power chain saws or grass trimmers. They are simple to use, reasonably priced, cheap to operate, and easy to service. Other stationary engines can be used to power such things as small compressors, light plants, grinders, and the like.

Essentially, the engine has but a single job—to provide power for operating various kinds of equipment. Yet in practice that equipment may vary considerably from turning a single blade attached directly to the engine crankshaft for cutting grass to driving a four- or five-speed transmission that provides various gear ratios for propelling a lawn or garden tractor. Or the engine may be utilized to work a simple pulley-and-belt arrangement, to move a sprocket that propels a moped by using a chain, or to drive an endless belt for operating a snowmobile. In addition the engine may not only propel the machine (such as a tractor), but also drive a working unit (such as a snow blower, tiller, or cutter). Components that are driven by the engine will be discussed separately in the latter part of the book.

Regardless of what the engine drives, it generally requires the most service and maintenance. Hence the engine maintenance and repair is covered in the first part of the book.

The book is limited to machines operated by single-cylinder gasoline engines, most of which develop less than 16 horsepower. The engine maintenance and repair sections cover both the two-stroke cycle and the four-stroke cycle engines.

TYPES OF SERVICE OPERATIONS

You might think that servicing all the different makes, models, and kinds of machines specifically mentioned in this book, as well as others that are not mentioned, would be a very difficult task. Yet if you take the time to thoroughly understand the fundamentals of operation and the correct approach to follow, you can efficiently perform the necessary preventative maintenance. You should also be able to make most of the needed repairs when problems, wear, or breakdowns occur. A complete understanding helps to avoid time-consuming errors and rework.

First, all small gasoline engines—whether two-stroke or four-stroke cycle—are made up of the same basic components. Certain things must take place before an engine can operate properly. If you understand why these things must happen and how they are made to happen, then you understand what is called the *component relationship* of the engine. More important, you will be able to identify malfunctions. While a certain amount of theory or *operating principles* is necessary, it is presented not for its importance as scientific principle but rather for its bearing on the parts or component relationship. Remember, if the engine is to operate, certain things must take place in a prescribed manner.

Secondly, small gasoline engines drive a large number of different machines. These include:

1. machines used to care for your lawn and/or garden,

2. chain saws for cutting wood,

3. vehicles for mobility and pleasure such as mopeds, golf carts, and snowmobiles, plus

4. a number of other machines used for pleasure or maintenance work, along with various specialized engine driven machines.

Usually the engine is serviced as a separate component, independently from the rest of the machine. The various kinds of equipment (machines) driven by the engine have many common elements. Figure 1-1 shows a typical four-stroke cycle engine used to power many different pieces of equipment.

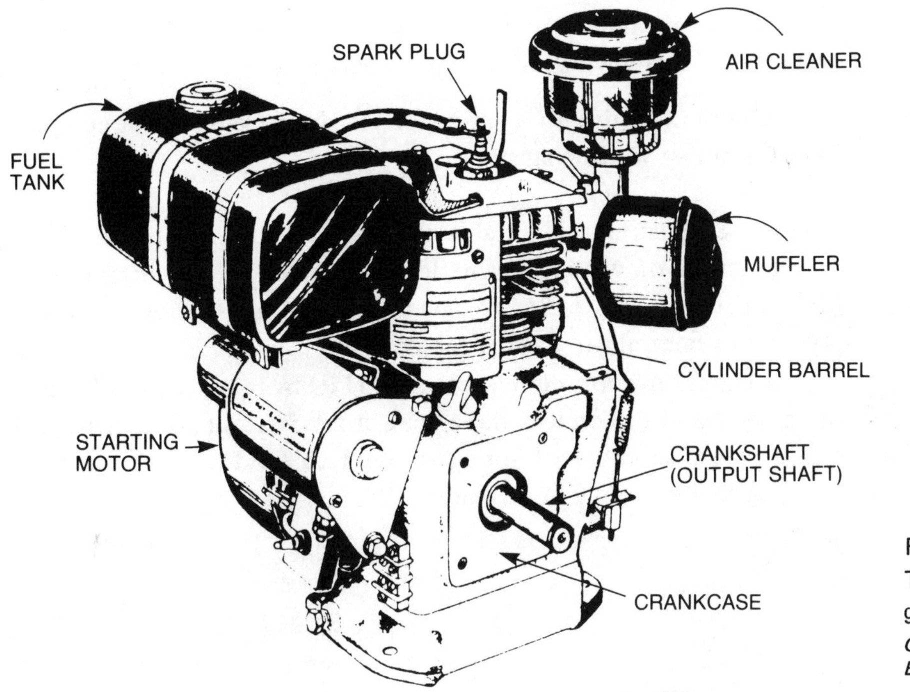

FIGURE 1-1
Typical four-stroke cycle gasoline engine
Courtesy Clinton Engines Corporation

COMPONENTS DRIVEN BY THE ENGINE

Most components driven by the engine are made up of some of the following units: gears, shafts, wheels, chains, sprockets, bearings, pulleys, belts, and linkages. When analyzing the equipment, you generally find that the engine drives the equipment through pulleys and belts, chains and sprockets, a variable sheave pulley or pulleys, a friction clutch, or a hydraulic transmission. These are all simple devices when viewed separately from the rest of the equipment.

CONNECTIONS BETWEEN POWER PLANT AND DRIVEN EQUIPMENT

Engagement means, simply, (connecting the engine to the equipment to be driven). In the case of a *belt and pulley*, engagement is generally a matter of tightening the belt so it turns the pulley attached to the equipment. Figure 1-2 shows a belt-and-pulley arrangement with a belt tightener to engage the belt. The shaft drives a sprocket.

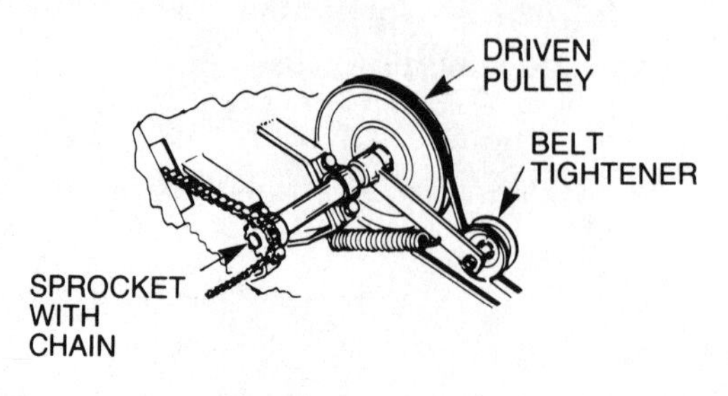

FIGURE 1-2
Belt tightner to apply power
Courtesy Tecumseh Products Co.

A *variable sheave pulley* may be used to connect the engine to the unit to be driven. This type of drive may be used in two ways:

1. as a single pulley with the engine mounted so as to maintain belt tension as the width of the pulley varies, or

2. in conjunction with a spring-loaded driven pulley that compensates for pulley width changes.

Figure 1-3 illustrates a variable drive sheave and a variable driven sheave, either of which automatically provides for varying ratios as the diameters of the pulleys change. On most installations they also provide for neutral (no driving force).

A simple *friction clutch* assembly (Figure 1-4) may also be used to connect and disconnect the engine from the unit to be driven. A spring or an assembly of springs is used to hold a friction surface (disc) against a turning drive plate.

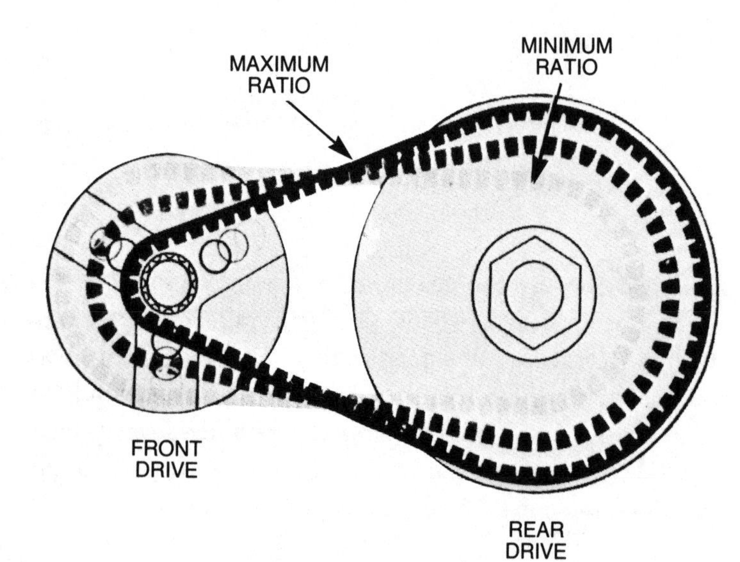

FIGURE 1-3
Drive and driven variable pulley sheaves which automatically provides varying driving ratios
Courtesy Harley-Davidson Motor Co., Inc. Subsidary of AMF Inc.

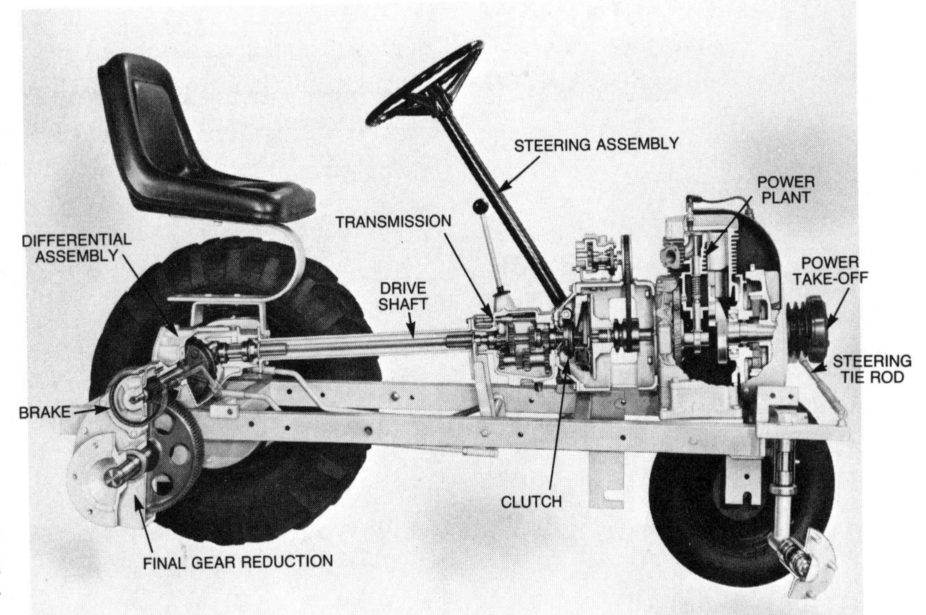

FIGURE 1-4
Cut-a-way view of a lawn & garden tractor
Courtesy Engineering Products Co.

CHANGING THE DRIVING FORCE

When gears are used for variable power development (*transmission*), as on some larger lawn and garden tractors, a disc clutch may be used to connect the engine to the transmission. The transmission drives the rear axle shafts through a differential assembly. The transmission permits the user to select different gear ratios to fit varying load requirements, as well as reverse.

A differential assembly at the rear driving axle permits one wheel to turn faster than the other when going around a corner.

Figure 1-4 is a cut-away illustration of a lawn and garden tractor showing the various components that transmit the driving force to the rear wheels.

A *centrifugal dog-type clutch* is generally used in a chain saw to apply driving force to the cutting chain. It automatically engages the cutting chain at a given speed.

The components used to transmit driving force may also be used to increase or to reduce power as well as speed. As an example, when a small pulley on the engine crankshaft is connected by a belt to a larger pulley, it causes the larger pulley to turn more slowly but with additional torque (turning power). The same principle applies when one sprocket drives another sprocket through a chain (moped). Differences in gear size bring about the same situation: A small gear driving a larger gear reduces the speed of the larger gear but increases the power output. Upon examination of the various machines, you find that this type of arrangement is always present in some form or another, unless the machine is a direct drive such as a lawn mower having the cutting blade attached directly to the engine crankshaft.

HOW PULLEYS, GEARS AND SHAFTS ARE SUPPORTED

Components such as gears, pulleys, sprockets, and the like are usually mounted on some type of shaft arrangement. The shaft is mounted on antifriction bearings or bushings to reduce wear and friction. Oil seals may be used on the shafts to keep the lubricant confined.

SERVICE OPERATIONS THE OWNER MAY PERFORM

When you look at a complete machine—such as a garden tractor, lawn mower, moped, or chain saw—you might get a feeling of uncertainty or a total lack of understanding. As you look back, however, this same frustration undoubtedly existed many times when first dealing with unfamiliar things. Yet as you study one system, unit, or component at a time and understand its function and relationship, your confidence will return and the task will become simple.

Operating principles are explained. These are important because they help you to see how each specific unit functions. Knowing what each component is supposed to do and how it should function helps you to determine whether or not it is operating properly. While differences exist from one machine to another, certain basic repair procedures are applicable to practically all units.

BECOMING FAMILIAR WITH THE VARIOUS COMPONENTS

Become thoroughly familiar with the machine you operate; learn about all the parts and how they relate to one another. Not only is this familiarity to your advantage from a maintenance and repair standpoint, but also as you become more familiar with your machine, you will be able to obtain maximum performance and service.

Illustrations of different makes and models of many machines are shown with the parts labeled. Use these figures to identify the components of your machine. Maybe they do not look exactly like the one shown in the illustration, but yours functions in much the same manner and may be serviced in the same way. Specific pictures are used to show the actual construction of the various parts.

Remember that the parts of the machine that fail or get out of adjustment do so by a gradual process. The process is speeded up by hard use and/or neglect, but it is slowed down by reasonable care and periodic maintenance. It is to your advantage to properly maintain the machine for care-free operation.

HOW THE BOOK IS STRUCTURED

At the beginning of each chapter a brief introduction describes why it is necessary to perform the different repair operations. This explanation gives you a better understanding of why certain repair operations are needed and the effect they have on operation.

Service operations are generally performed for the following reasons:

1. The machine no longer operates.

2. The machine operates but not in a satisfactory manner—the engine misses or runs rough.

3. There is excessive noise.

4. The engine is hard to start.

5. The engine lacks power.

6. There is obvious wear and/or excessive free play in some particular part.

7. The machine needs adjustment because it is not operating properly.

Emphasis is placed on practical maintenance and repair procedures. The purpose is to keep the machine operating at the performance level originally engineered into the unit.

Only a few tools and minimal equipment are needed for the majority of jobs. Most of the tools are the common type that many people have if they make repairs around the home.

The more common maintenance operations—such as changing oil, lubricating, adjusting and/or replacing belts or chains, tuning the engine, cleaning the air filter, and making other adjustments—can be performed by most anyone.

In most cases an owner/operator's manual comes with a new machine. This manual generally indicates when and what services should be regularly performed. It also gives operational specifications, such as spark plug gap setting, ignition contact point setting, and so on. This booklet is very helpful in servicing a machine.

LOCATING TROUBLES

When malfunctions occur, often the most difficult task is to decide what is wrong. By carefully looking at the different units, you may find that you can do some repairs without detailed instructions—such as changing a spark plug, replacing a belt, or changing oil. Some jobs, however, cannot be done without special equipment—such as honing or reboring a cylinder, refacing valves, or fitting a piston pin.

By using a systematic approach and an elimination process, you can generally locate the area where the problem exists. As an example, a lawn mower has "conked out" on you. You remove the wire from the spark plug, hold the terminal end about a quarter-inch from the ground (the cylinder head), and crank the engine. You see that a spark occurs at the gap between the terminal and ground, so you know that the ignition system is OK. You then look elsewhere for the trouble.

In general, for an engine to run, three conditions must exist:

1. It must develop enough compression (pressure) to compress the air-fuel mixture in the combustion chamber.

2. A properly proportioned air-fuel mixture must be reaching the combustion chamber.

3. A spark with a high enough voltage to jump across the spark plug gap must be delivered within the combustion chamber at the exact time to ignite the air-fuel mixture.

Before you undertake any type of service operation, you should take the following steps:

1. Evaluate the problem to determine what needs to be done.

2. Decide whether you should tackle it yourself or let a professional do it.

3. If you think you can do the job, decide whether you have the necessary parts, tools, and time.

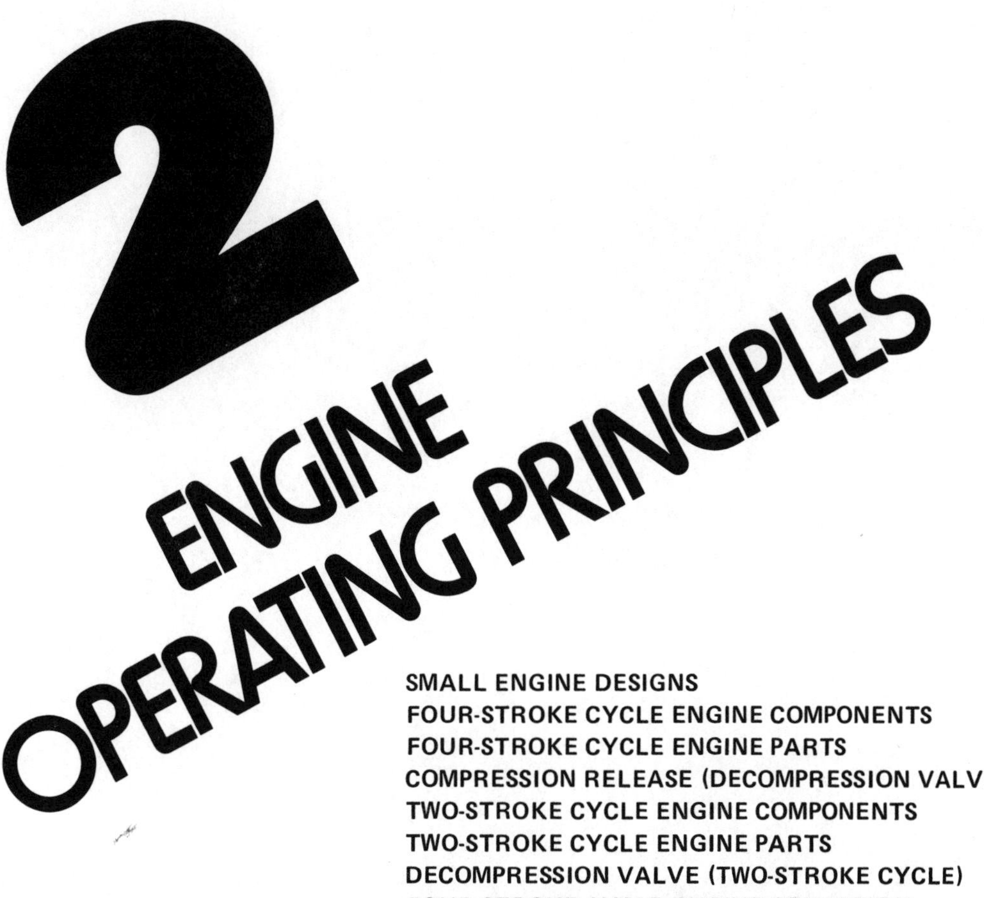

SMALL ENGINE DESIGNS
FOUR-STROKE CYCLE ENGINE COMPONENTS
FOUR-STROKE CYCLE ENGINE PARTS
COMPRESSION RELEASE (DECOMPRESSION VALVE)
TWO-STROKE CYCLE ENGINE COMPONENTS
TWO-STROKE CYCLE ENGINE PARTS
DECOMPRESSION VALVE (TWO-STROKE CYCLE)
FOUR-STROKE CYCLE ENGINE OPERATION
TWO-STROKE CYCLE ENGINE OPERATION
CHARACTERISTICS OF THE TWO-STROKE CYCLE ENGINE

The beginning of the chapter includes the functions of the various components that make up different engines. Being able to recognize the different parts and to understand their relationship goes a long way toward developing an understanding of operating principles.

The four-stroke cycle and two-stroke cycle principles of operation follow. This section on operating principles makes you aware of what must happen within an engine for it to operate. Also, by developing an understanding of operating principles, you should be better able to isolate engine troubles. You should know why it is important for the rings to seal properly, for the valves to seat without leaking, for gaskets to seal properly, for the spark to occur at exactly the right time, and for the proper air-fuel mixture to be available when required. Also included are general factors that have a bearing on engine operation. Later chapters cover the specific service procedure for each component.

SMALL ENGINE DESIGNS

Many different makes and models of small gasoline engines are in general use in many different applications. Yet all small gasoline engines are of the internal combustion design, they depend on air, fuel, and ignition, and they operate on either the two-stroke or the four-stroke cycle principle of operation.

In both types of engines an air-fuel mixture must be drawn into the cylinder, compressed, and ignited; and the remaining unburned gases must be expelled. These operations are carried out, according to engine design, by means of:

1. *A two-stroke cycle:* two strokes of the piston (one up and one down) with one crankshaft revolution *or*

2. *A four-stroke cycle:* four strokes (two up and two down) of the piston with two revolutions of the crankshaft.

All engines must be designed and constructed so as to bring about these conditions. Both types of engines are known as "heat engines" because they convert the heat energy contained in the fuel into mechanical energy. They may be further classified as "internal combustion engines" because the fuel is burned inside the engine cylinder.

From an operator's standpoint, the major difference between these two principles is that, with a two-stroke cycle engine, the lubricating oil must be mixed with the gasoline. The four-stroke cycle engine has a separate fuel tank for the gasoline supply, and the lubricating oil is put directly into the engine; the oil is not burned with the fuel.

All engines, regardless of type, must have certain basic components to operate. Despite differences in size and shape depending on the application of the power plant, the servicing is much the same regardless of these differences.

FOUR-STROKE CYCLE ENGINE COMPONENTS

A typical single-cylinder four-stroke cycle engine, with the components labeled, is shown in Figure 2-1. This is a *horizontal engine.* The crankshaft is located in the engine in a horizontal position with the cylinder and piston in a vertical position. The power is taken off the side of the engine. In a *vertical engine,* the crankshaft is in a vertical position with the power take-off at the bottom of the engine. Become familiar with the names and location of the parts in the illustration, so you can identify them in the engine you are interested in servicing.

After becoming familiar with the various parts and their relationships, understanding what takes place to bring about the four-

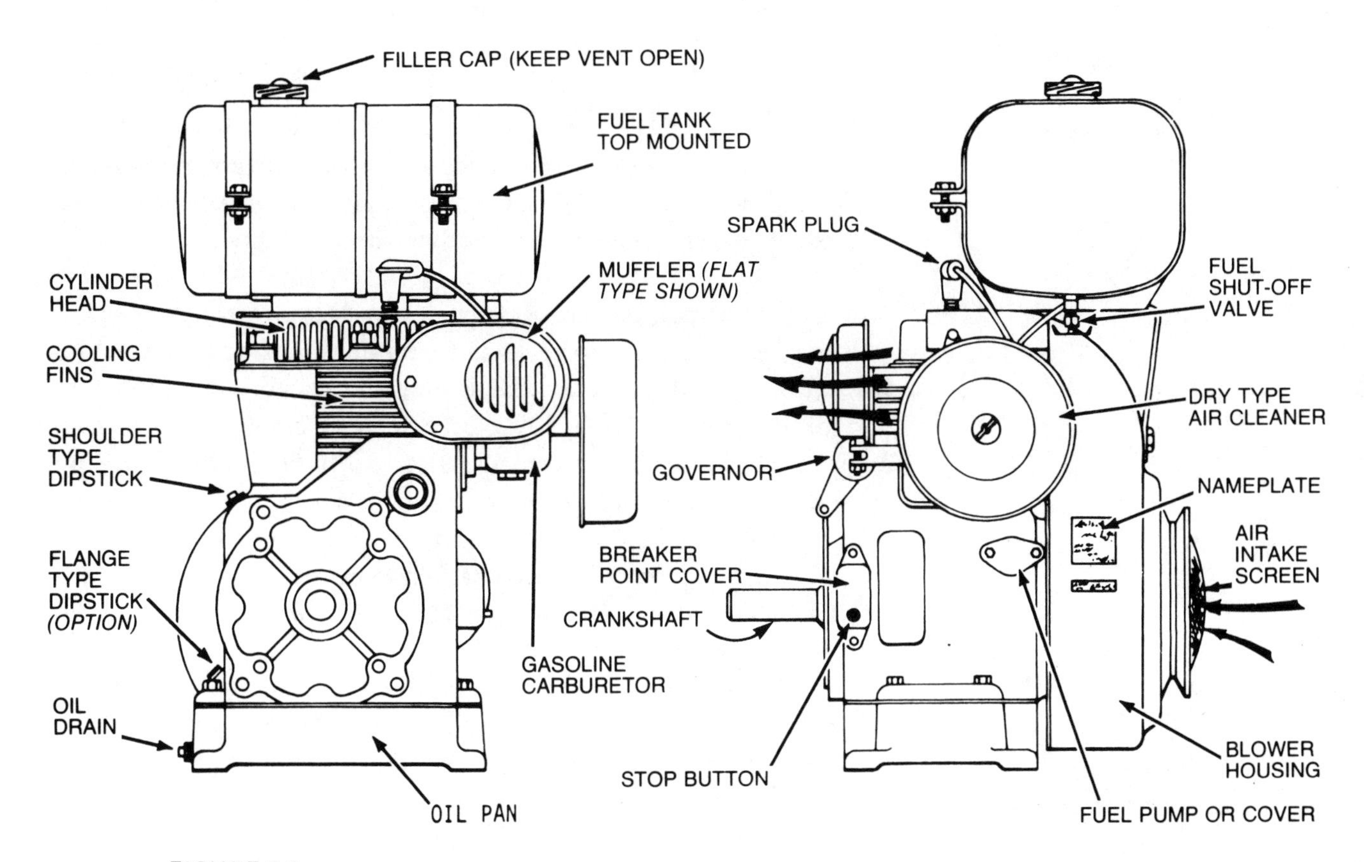

FIGURE 2-1
Typical four-stroke cycle single-cylinder engine
Courtesy Kohler Co.

stroke cycle operation is a simple matter. Once you understand the basic principle of operation, diagnosing troubles and making repairs are comparatively easy matters. Clearance between parts, sequence of events, timing, lubrication, air-fuel mixture, and ignition all become meaningful from a service standpoint. The specifics relative to construction features, maintenance, trouble shooting, replacement, and repair of each unit are covered in later chapters.

FOUR-STROKE CYCLE ENGINE PARTS

Let's review the parts quickly before getting into any great detail:

1. A typical one-cylinder engine consists of a *cylinder assembly* made up of a cylinder head and cylinder barrel, which contains a piston and in most cases two valves. The assembly is either bolted to the crankcase or cast as one unit with the main section of the crankcase.

2. A *crankshaft* is supported on bearings in the crankcase housing. Also located in the crankcase, is the valve-operating mechanism along with some system for lubricating the moving parts. A connecting rod connects the piston to the crankshaft.

3. A *carburetor* is mounted on the engine with a passageway in the cylinder between the carburetor and intake valve port.

4. An *exhaust system* is used to carry the unburned gases from the combustion chamber to the outside.

5. The *ignition system*, mounted on the side of the crankcase, is driven by the camshaft. The flywheel is mounted on one end of the crankshaft, which makes for a smoother running engine.

Now let's look at these parts more closely.

The cylinder assembly, sometimes referred to as the "cylinder block," is a metal container that is long and round like a pipe. One end of the cylinder is tightly closed with a removable cover (the cylinder head), while the other end is open. Metal fins normally extend around the outside of the cylinder assembly and across the cylinder head to cool the engine by disipating the heat. A metal shroud (or cover) may be fitted around the cylinder to better distribute air around the fins. Figure 2-2 is a cut-away view of a typical single-cylinder engine.

Acting as a plug, a piston is fitted snugly into the open end of the cylinder. Flexible piston rings are installed in grooves in the piston to form a sliding seal between the piston and cylinder wall.

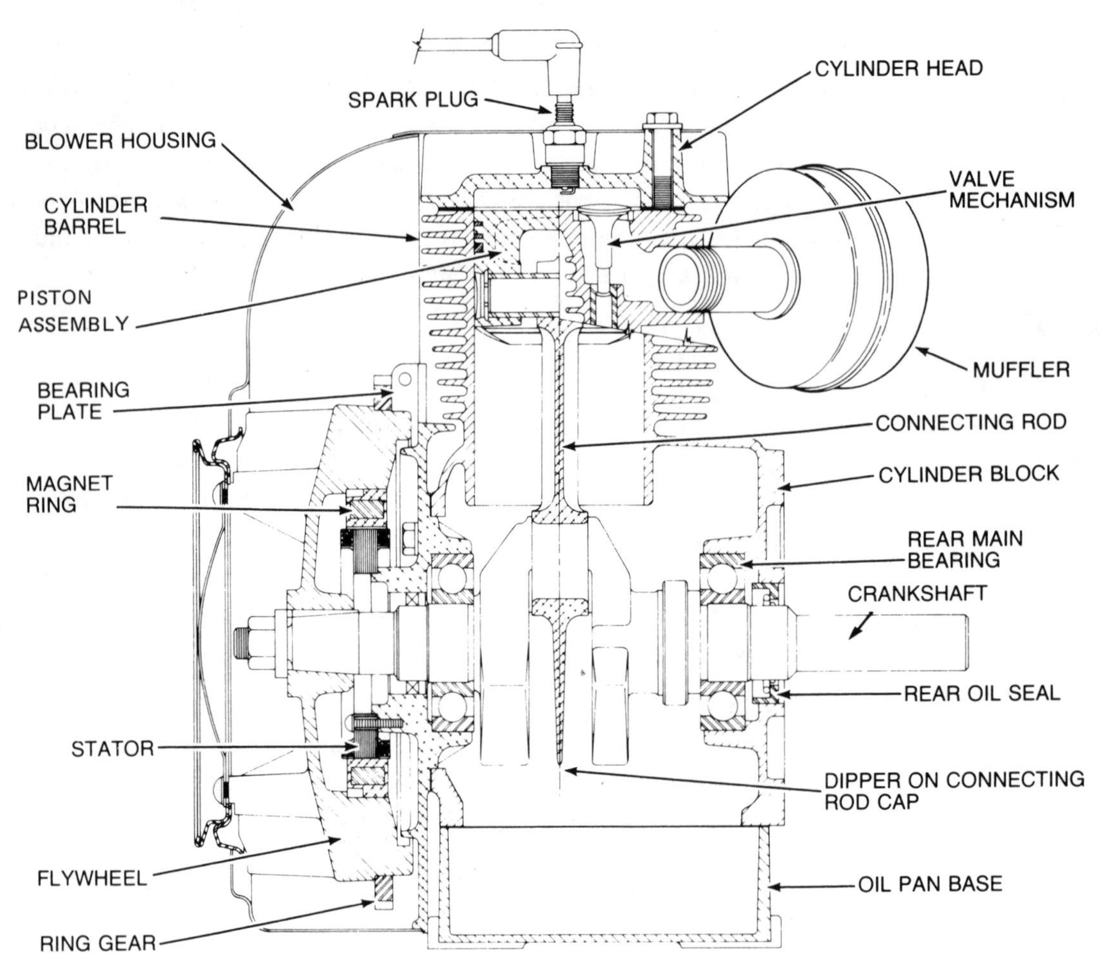

FIGURE 2-2
Cut-a-way view of a typical single-cylinder engine
Courtesy Kohler Co.

As the piston moves up and down in the cylinder, it increases and decreases the space above the piston.

The cylinder assembly may be bolted to a crankcase, or both the cylinder and the crankcase may be cast as a single unit with removable end sections. The crankcase serves as a base for the engine.

The crankshaft is supported on antifriction bearings in the crankcase.

When an air-fuel mixture is drawn into the space above the piston, compressed and burned, pressure from the expanding gases push the piston downward. To harness this force, some type of mechanical linkage must change the up-and-down movement of the piston to a rotary or turning motion. To do so, a connecting rod, attached to a piston pin inside the piston, connects the piston to the crankshaft. The piston pin acts as a pivot to allow the connecting rod to swing from side to side like the pendulum in a clock. This pendulum action permits the piston to be attached to the circularly turning crankshaft. The lower end of the connecting rod is attached to the crankshaft journal. The journal is positioned on the crankshaft in such a manner that it is offset from the pivot point (the bearing) of the crankshaft. This offset, called the "crank," provides the leverage needed by the explosion to turn the crankshaft, thus converting the up-and-down motion of the piston into a rotary motion. This effect is known as "reciprocating motion."

The engine must also provide some means of getting an air-fuel mixture into the combustion chamber, which is the space between the top of the piston and the cylinder head when the piston is at top dead center (TDC). (A certain amount of space, depending on the compression ratio desired, is usually cast into the cylinder head for this purpose.) To provide a passage for the mixture between the carburetor and combustion chamber, two valve ports (openings) are located alongside and at the top of the cylinder. One port is connected to the carburetor, and the other is connected to the exhaust system. Valves are fitted over the port openings to open and close the ports. A camshaft arrangement, driven by the crankshaft, controls the opening and closing of the valves. This arrangement permits an air-fuel mixture to be drawn into the cylinder, compressed and burned. The burned gases are then expelled after combustion.

A carburetor is used to mix gasoline and air, as well as to supply the mixture to the engine in the correct ratio. It also controls the amount of air-fuel mixture entering the engine, which affects engine speed. To obtain power from the fuel mixture, it must be "squeezed" (or compressed): the more it is compressed, the more power you obtain, up to a limit. In other words, higher compression ratios mean greater power output. An ideal air-fuel mixture is generally considered to be approximately fifteen parts of air to one part of gasoline by weight.

Fuel may be delivered to the carburetor by a gravity feed system. With this type of system, the fuel supply tank must be located above the carburetor. Alternatively, a few engines use a fuel pump, which is necessary when the fuel supply is located below the carburetor. The fuel pump is of the diaphragm type. The purpose of the air cleaner, installed on the intake of the carburetor, is to filter the incoming air before it enters the carburetor. A governor is generally attached to the carburetor throttle through a linkage arrangement. The purpose of the governor is to help maintain a constant engine speed according to the throttle setting.

After the fuel mixture has been compressed, it must be ignited and thus caused to expand. Accordingly, the ignition system creates a high-voltage spark. This spark must be provided at exactly the right time, and it must be strong enough to jump across the gap between the spark plug electrodes. The spark plug is normally installed in the cylinder head with the electrodes extending into the combustion chamber. Either a battery ignition system or a magneto may develop this high-voltage spark.

All moving parts must be lubricated to reduce friction and heat, as well as to help seal some parts. To lubricate the engine, some machines use an oil slinger, which is a finned plate that revolves when the engine is running and that distributes the oil inside the engine in the form of a spray. A dipper may be used on the connecting rod cap to make sure the connecting rod bearing is lubricated. A few engines are equipped with an oil pump to assure positive oil delivery to the moving parts.

A flywheel is attached to one end of the crankshaft. It is heavy because it provides the intertia for the crankshaft that is necessary to smooth out engine operation. If a magneto type of ignition system is used, the flywheel generally contains magnets and is considered part of the magneto.

Counterweights are also used on the crankshaft to assist in bringing about smoother engine operation. With just one power impulse every two revolutions (one revolution with a two-stroke cycle engine) of the crankshaft, inertia tends to drop off. Weights that are properly located on the crank arms help to carry the inertia over for smoother operation. Some engines have counterweights that are geared to rotate in one direction or the other so as to shift the weight for less vibration as the crankshaft turns.

COMPRESSION RELEASE (DECOMPRESSION VALVE)

Some small engines, particularly those which higher horse power ratings, may have a compression release mechanism to reduce compression during starting. This enables the engine to turn over more easily when starting. Different types of compression release mechanisms are found on the various engines.

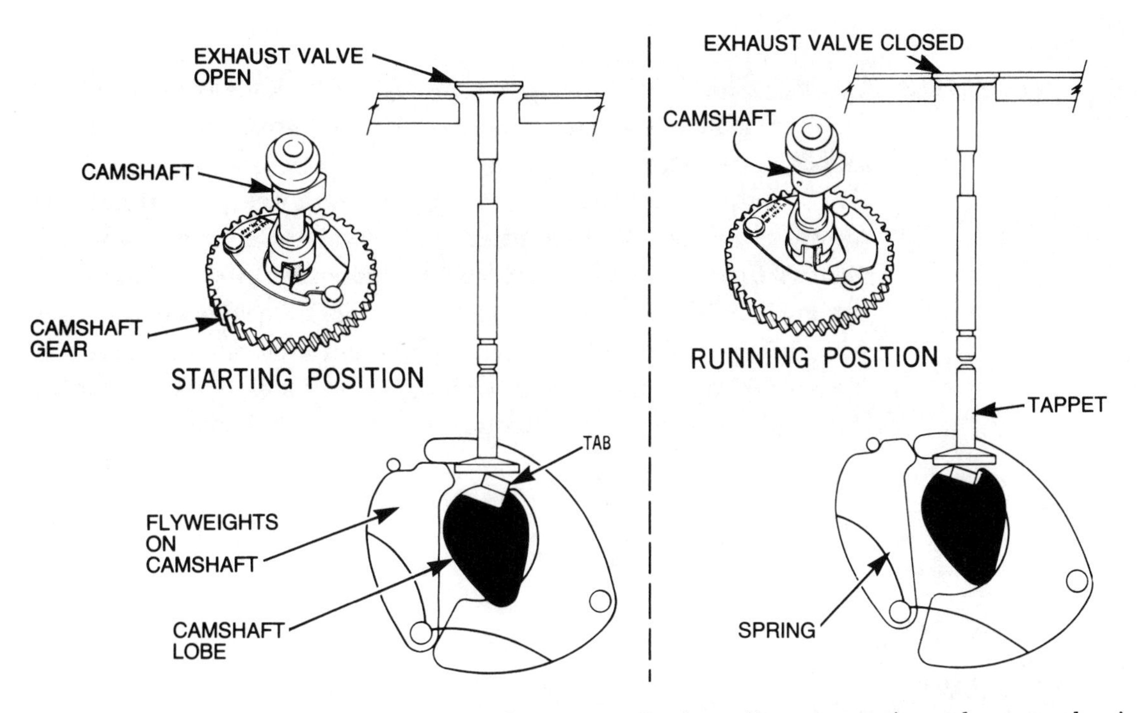

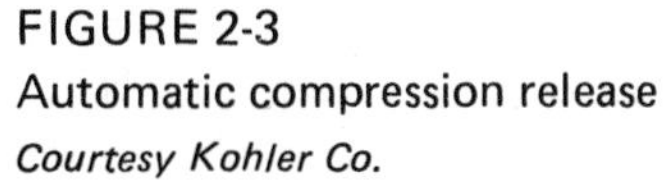
FIGURE 2-3
Automatic compression release
Courtesy Kohler Co.

One type of automatic compression release mechanism is shown in Figure 2-3. On this installation the exhaust valve lifts off its seat during cranking and then reacts to resume full compression as soon as the engine reaches operating speed. This action is accomplished by means of flyweights on the cam gear. The flyweights are held in the inner position by spring tension during starting. The tab on the larger flyweight protrudes above the profile of the exhaust cam. This tab lifts the exhaust valve off its seat during the first part of the compression stroke, thus releasing compression. As soon as the engine starts, centrifugal force moves the flyweights to the outer position. This position allows the tab to drop into a recess in the exhaust cam so the exhaust valve can operate in a normal manner.

Some engines that do not have a compression release instead have an automatic spark retard to prevent the engine from kicking back. This retardation makes for easier starting. The ignition timing is retarded during starting and gradually advances after the engine has started.

Another method of reducing compression while starting, called "Easy Spin," is used on some small engines. The intake lobe on the camshaft is ground with a small ramp, which holds the intake valve open one-hundredth of an inch for a small fraction of a section of the compression stroke. At slow cranking speed this interval of time is long enought to let enough pressure in the combustion chamber escape so as to reduce compression. Again, the engine is cranked over more easily. At operating speeds the time interval is so short that there is practically no noticeable escape of pressure and so the horse power is not impaired. With an arrangement of this type, you must turn the engine flywheel backward when checking compression.

TWO-STROKE CYCLE ENGINE COMPONENTS

The two-stroke cycle engine uses many of the same components as the four-stroke cycle engine. So a lot of the information is the same as for the four-stroke cycle engine.

A typical two-stroke cycle engine, with the parts labeled, is shown in Figure 2-4. Become familiar with these parts so you can identify them with the engine you are interested in servicing. By becoming familiar with the various parts, with their operation, and with their relationships, you will understand what must take place if the engine is to function properly. Detailed descriptions, trouble shooting, and servicing are all discussed in later chapters.

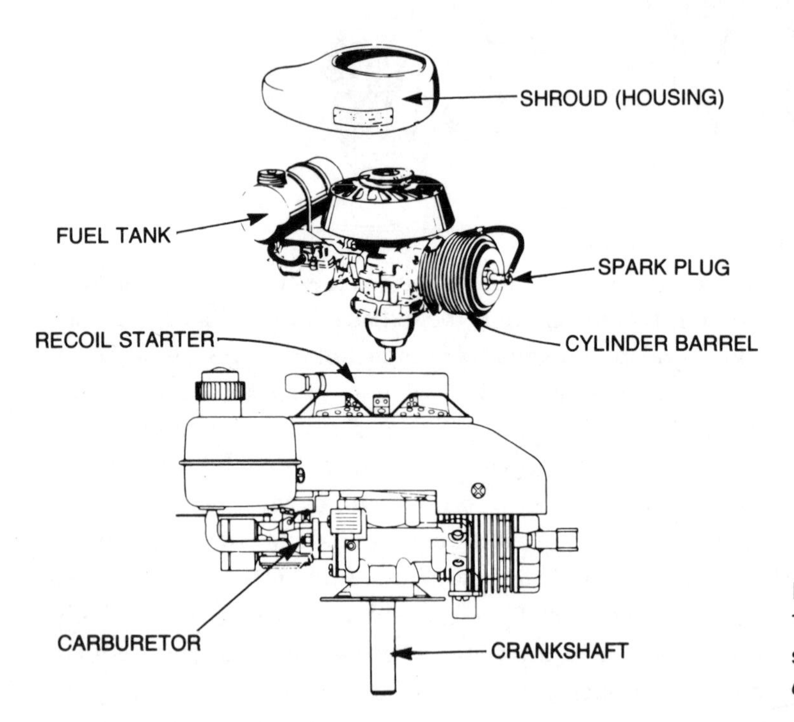

FIGURE 2-4
Typical two-stroke cycle single-cylinder engine
Courtesy Tecumseh Products Co.

TWO-STROKE CYCLE ENGINE PARTS

This type of engine consists of two major components: the cylinder assembly and the crankcase assembly (Figure 2-5). The cylinder assembly is sometimes called the "cylinder block" or "cylinder barrel." The cylinder itself is a metal container that is long and round like a pipe. The cylinder's outside surface is finned for cooling purposes. One end of it is tightly closed with a removable cylinder head, which is also finned for cooling purposes. The other end is open. Two ports (openings) are located in the cylinder wall. Fuel enters the cylinder through one port (the intake), and exhaust gases are ejected through the other port (the exhaust). Either the cylinder assembly may be bolted to the crankcase assembly, or it may be cast as one unit with the crankcase.

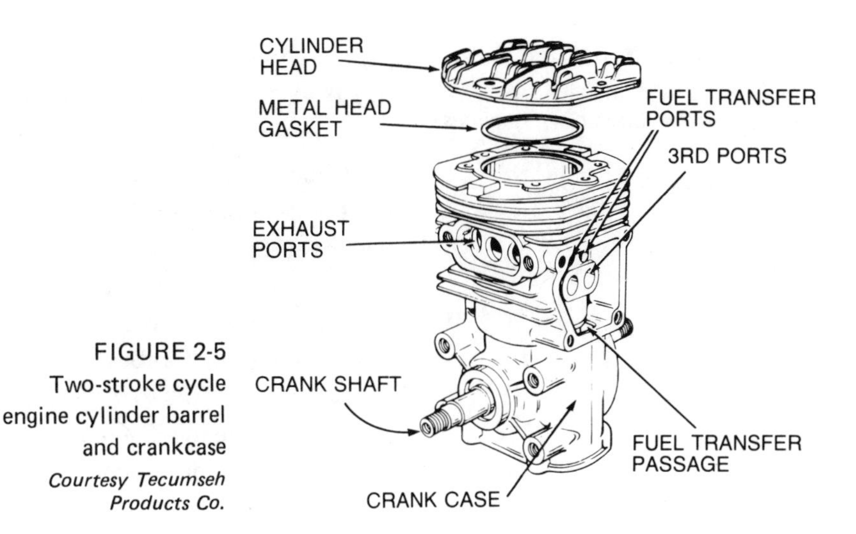

FIGURE 2-5
Two-stroke cycle engine cylinder barrel and crankcase
Courtesy Tecumseh Products Co.

Acting as a plug, a piston is fitted snugly into the open end of the cylinder. Flexible piston rings are installed in grooves in the piston to form a sliding seal between the piston and cylinder wall. As the piston moves up and down in the cylinder, it increases and decreases the space between the top of the piston and the cylinder head. The combustion chamber is the space between the top of the piston and the cylinder head when the piston is at top of its travel (TDC). A certain amount of space, depending on the compression ratio desired, is usually cast into the cylinder head for this purpose.

The carburetor is mounted on or connected to the crankcase. Since the crankcase is used to compress the air-fuel mixture, it must be air tight. So on most engines some type of one-way valve is inserted between the carburetor and the crankcase inlet opening. This valve permits the air-fuel mixture to be drawn into the crankcase, but it prevents the mixture from being forced out of the crankcase as it is being compressed. The type of valve may vary, but in most cases it is a reed valve, which is opened by vacuum (suction) and closed by pressure.

The crankcase serves as a base for the engine. The crankshaft is supported on antifriction bearings in the crankcase. When an air-fuel mixture is admitted into the space above the piston, it is compressed and burned. The resultant pressure from the expanding gases forces the piston downward. To harness this force, some type of mechanical linkage must change the up-and-down movement of the piston to a rotary or turning motion. To do so, a connecting rod, attached to a piston pin inside the piston, links the piston to the crankshaft. This pin acts as a pivot to allow the connecting rod to swing from side to side like the pendulum in a clock. This pendulum action permits the piston to be attached to the circularly turning crankshaft. The lower end of the connecting rod is attached to the crankshaft journal. The journal is positioned on the crankshaft so that it is offset from the pivot point (bearing) of the crankshaft. This offset, called the "crank" or "crank throw," provides the leverage needed by the explosion to turn the crankshaft—thus converting the up-and-down motion of the piston into rotary motion.

Counterweights are generally a part of the crank throws to provide for a smoother-running engine. This additional weight helps to carry over the inertia of the power impulse, which lasts for only a small part of the crankshaft revolution.

When you get to the means for getting the air-fuel mixture into the combustion chamber, you see where the two-stroke cycle engine differs from the four-stroke type. Instead of an intake valve, the two-stroke cycle engine has an intake port opening on one side of the cylinder. An exhaust port opening is located on the opposite side of the cylinder. The carburetor is mounted on the crankcase, which acts as a receiver for the air-fuel mixture. An intake bypass (passageway) leads from the crankcase to the intake port. A one-way valve, generally a reed valve located between the carburetor and crankcase, permits the air-fuel mixture to enter the crankcase, but it also prevents the mixture from being forced out through the carburetor as pressure builds up in the crankcase. The valve arrangement may vary with engine design.

The type of camshaft and valve arrangement used in a four-stroke cycle engine is unnecessary in a two-stroke cycle engine.

A carburetor mounted on the crankcase mixes gasoline and air and supplies the mixture to the engine crankcase in the proper ratio. The engine speed is controlled by a throttle arrangement in the carburetor. A governor, generally connected to the carburetor throttle vale, aids in maintaining a fixed engine speed regardless of the load factor.

To get maximum power from the fuel, the mixture of gasoline and air must be compressed: the more the mixture is squeezed (compressed), the more power you obtain within certain limits. A higher compression ratio means that the engine is capable of developing more power.

After the air-fuel mixture has been compressed, it must be ignited. The rapid burning of the fuel (the explosion) forces the piston downward, thus creating power. The ignition system brings about this explosion by developing a high-voltage spark. The spark must be delivered at exactly the right time, and it must be strong enough to jump across the gap between the electrodes of the spark plug. The spark plug is installed in the cylinder head with the electrodes extending into the combustion chamber. Either a battery ignition system or a magneto is used to develop the necessary voltage to fire the spark plug.

Since the lubricating oil is mixed with the gasoline, no separate lubrication system is needed.

Attached to one end of the crankshaft is a flywheel, which is a heavy wheel used to provide intertia to smooth out the firing impulses and thus make for a smoother-running engine. When a mageto ignition system is used, the flywheel generally contains magnets and is considered part of the magneto operation.

DECOMPRESSION VALVE (TWO-STROKE CYCLE)

A decompression valve or compression release valve, used on some engines, allows the operator to open a valve in the combustion chamber to release some of the compression when starting. The reason: Pulling the starter rope (to turn the engine over) may be hard, particularly with larger engines.

In the two-stroke cycle engine, the decompression valve is usually a small spring-loaded poppet-type valve similar to the valves used in a four-stroke cycle engine. It may be opened by a cable-and-lever arrangement or simply by pressing a button, either of which forces the valve open *against* the spring tension. The key characteristic is that the valve is opened against spring tension before cranking the engine. Some valves automatically close when the engine starts, and others must be closed manually.

FOUR-STROKE CYCLE ENGINE OPERATION

For any engine to operate, a series of events must take place in a specific order. Of the two different types of engines, the four-stroke cycle engine is the more popular and has the wider usage. In this type of engine, a power cycle includes four events inside the cylinder within two revolutions of the crankshaft: (1) intake, (2) compression, (3) power, and (4) exhaust. Figure 2-6 illustrates the four strokes.

To better understand how the engine operates, let's start with the piston in its uppermost position, that is, at top dead center (TDC). On the intake stroke, the piston moves downward in the cylinder. Because the piston fits tightly in the cylinder, it creates a vacuum above itself as it moves down. As the piston starts down, the intake valve opens, and the exhaust valve is closed. As the intake valve opens, it uncovers the intake port which is connected to the carburetor. The vacuum created by the downward piston movement

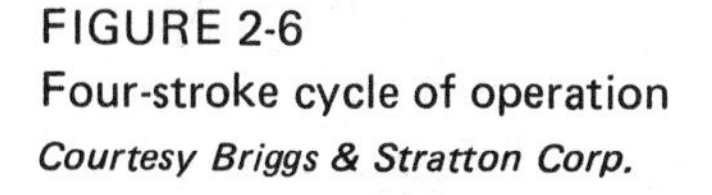
FIGURE 2-6
Four-stroke cycle of operation
Courtesy Briggs & Stratton Corp.

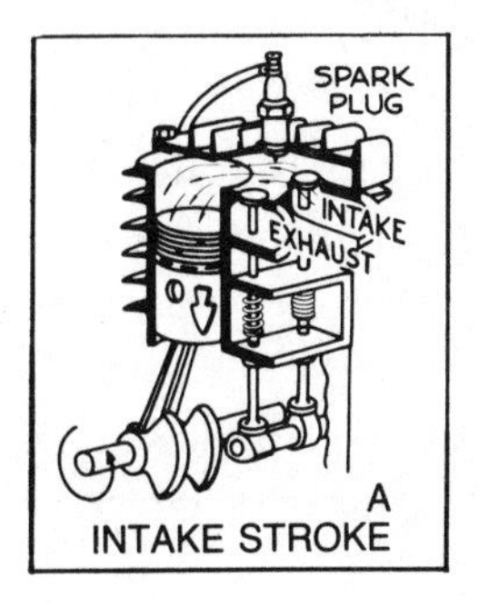

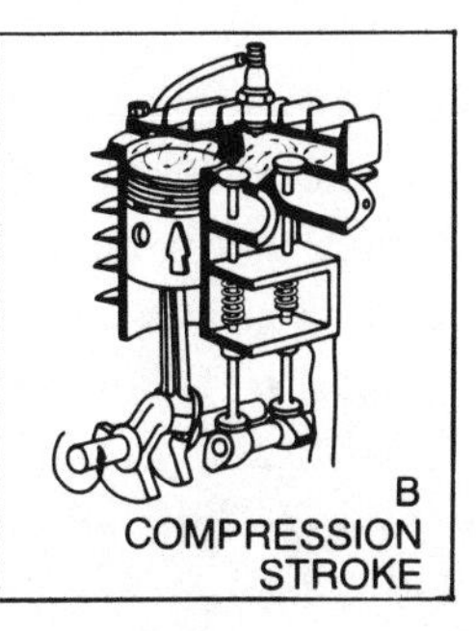

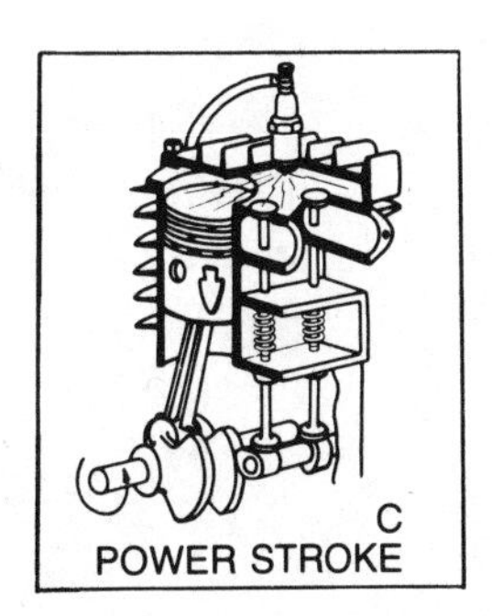

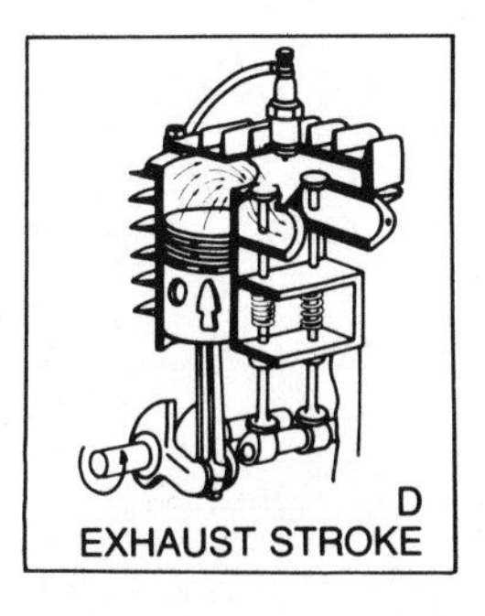

sucks air through the carburetor. As air moves through the carburetor, it picks up gasoline which becomes a vapor. The air-fuel vapor fills the space within the cylinder above the piston as the piston continues to move down. When the piston reaches the bottom of its travel, bottom dead center (BDC), it starts upward. The distance that the piston moves up and down—known as the "stroke"—is controlled by the length of the crank arm on the crankshaft. As the piston moves upward again, the intake valve closes and the air-fuel mixture is compressed (squeezed). This is the compression stroke.

The amount of compression depends on the compression ratio. This is the ratio between the volume in the cylinder when the piston is at BDC and the volume when the piston is at TDC. The compression pressure is generally from 100 to 150 pounds per square inch (psi). As the piston reaches TDC the spark plug fires igniting the mixture. Immediately the pressure in the combustion chamber rises to nearly 600 - 700 psi. With a piston 3 1/2 inches in diameter, the pressure is close to 3 tons on the top of the piston. This pressure forces the piston down, turning the crankshaft and whatever is connected to the shaft. This is the power stroke.

Both valves are closed during the compression stroke and the power stroke. At the bottom of the power stroke the exhaust valve opens. When the piston moves up it forces the exhaust gases out through the exhaust valve, through the exhaust port, and into the exhaust pipe and muffler. This is the exhaust stroke. As the piston gets near the top of the exhaust stroke, the exhaust valve closes and the intake valve opens—ready for the start of the next cycle.

Since the crankshaft goes around twice during each cycle, only one stroke out of four delivers power. During the other three strokes the crankshaft is acting on the piston, pushing it up and pulling it down. The heavy flywheel attached to the crankshaft, along with the offset counterweights on the crankshaft, aids in keeping the crankshaft turning smoothly between power strokes.

TWO-STROKE CYCLE ENGINE OPERATION

The name "two-stroke cycle engine" gives you an idea of how it compares with the four-stroke cycle. Instead of going through four strokes of the piston—two up and two down—before repeating, each cycle has only two strokes—one up and one down. The cylinder fires once for each revolution of the crankshaft. The intake and compression strokes are combined, as are the power and exhaust strokes. This type of power cycle permits the engine to produce a power stroke for every two piston strokes or for every crankshaft revolution. Figure 2-7 illustrates the events as they take place within the two-stroke cycle engine.

Two-stroke cycle engines normally have intake and exhaust port openings on the side of the cylinder rather than valves (Figure 2-5).

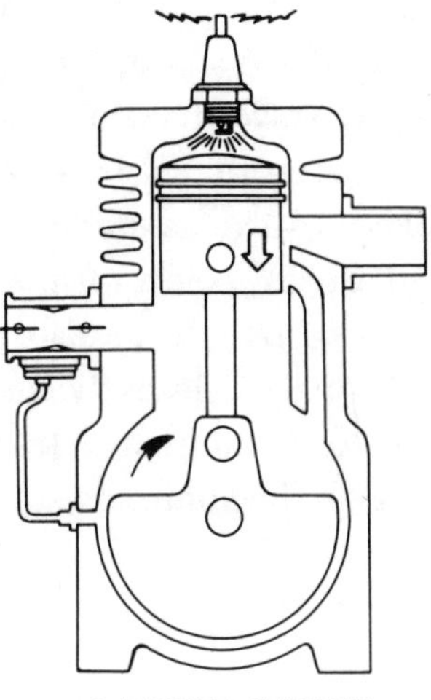

1. POWER STROKE

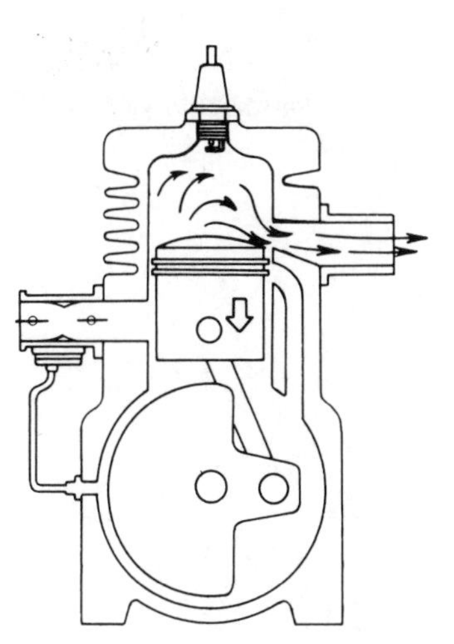

2. POWER STROKE & EXHAUST

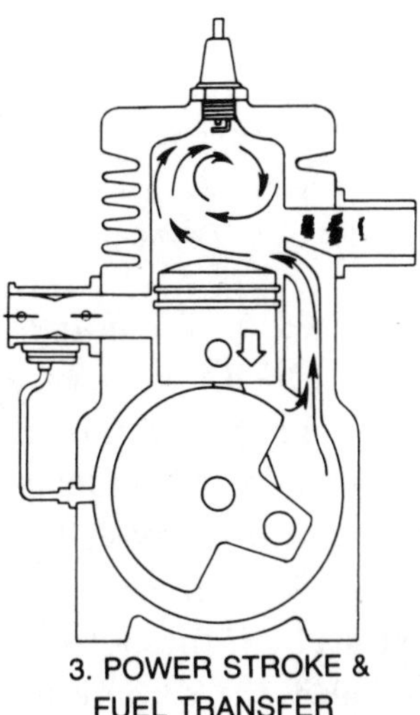

3. POWER STROKE & FUEL TRANSFER

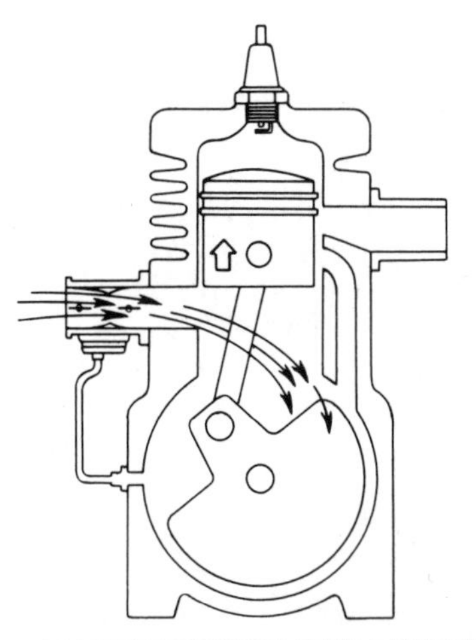

4. COMPRESSION STROKE & INTAKE

FIGURE 2-7
Two-stroke cycle of operation
Courtesy Deere & Company, Moline, Illinois

The air-fuel mixture is admitted from the carburetor into the crankcase and then into the combustion chamber through the intake port. Some engines have a one-way valve between the crankcase and carburetor. The exhaust port is connected to the exhaust system, which carries away the unburned gases.

There are two types of two-stroke cycle engines: (1) the piston-ported engine and (2) the engine using a one-way valve.

The piston-ported engine is the more popular one. This type has a cylinder barrel with three ports (or holes) in the cylinder wall: an inlet port, a transfer port, and an exhaust port (Figure 2-5). The inlet port admits the air-fuel mixture from the carburetor into the crankcase. The transfer port is connected to a passageway from the crankcase to the cylinder. The exhaust port is connected to the exhaust system to conduct the unburned gases to the outside. The ports are opened (uncovered) and closed (covered) by the movement of the piston.

While the piston-ported engine has three ports, the reed valve engine uses only two: a transfer port and an exhaust port. The reed valve, a one-way valve that consists of a spring steel plate (the reed), is mounted between the carburetor and crankcase. Figure 2-8 shows some typical reed valve assemblies. The valve opens when the crankcase pressure drops below atmospheric pressure (thus creating a vacuum). The air-fuel mixture is drawn from the carburetor into the crankcase. The valve closes when the pressure reverses, (and the vacuum is no longer present), trapping the air-fuel mixture in the crankcase so it can be compressed.

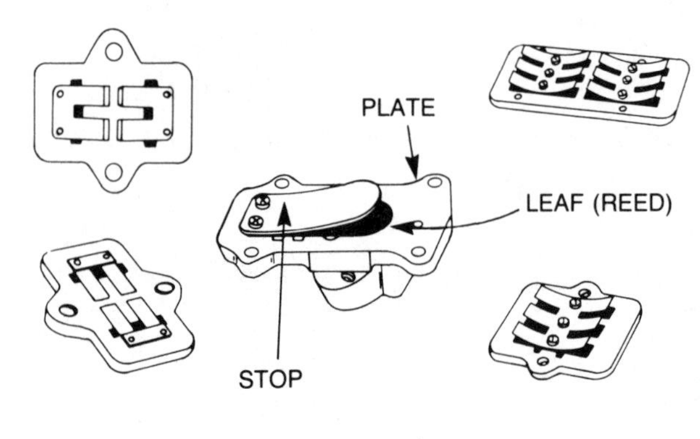

FIGURE 2-8
Typical reed valve assemblies
Courtesy Tecumseh Products Co.

At low speed, piston-ported engines tend to backfire through the carburetor, which upsets the air-fuel mixture. But, since the reed valve keeps the mixture in the crankcase, it can develop more power at low speeds because more fuel gets into the engine.

How do you distinguish between the two engines? On piston-ported engines, the carburetor is positioned on an intake manifold (a tube) that leads to the bottom of the cylinder barrel. On engines using a reed valve, the carburetor is mounted directly on the crankcase.

To understand the two-stroke engine operation, let's start with the piston at the bottom of its travel—that is, at bottom dead center (BDC). When the crankshaft is turned, the piston moves from bottom dead center to top dead center (TDC), which is as far up as the piston can travel. Because the piston fits tightly in the cylinder and because the crankcase is sealed, a vacuum is created in the crankcase below the piston when the piston moves upward. As the piston nears the end of its stroke the intake port in the cylinder is uncovered. Since the port is connected to the carburetor, the vacuum in the crankcase draws the air-fuel mixture from the carburetor into the crankcase. When a one-way reed valve is used between the carburetor and crankcase, the valve opens as the vacuum builds up in the crankcase. Thus the air-fuel mixture enters from the carburetor.

When the piston reaches the end of its upward motion (TDC), it reverses direction and starts downward. The piston now covers the intake port; or, if a valve is used, the valve is closed by spring action because the vacuum is no longer present. The crankcase is now sealed so that additional downward travel of the piston compresses the air-fuel mixture in the crankcase. Near the bottom of the piston travel, an intake bypass port leading from the crankcase to the combustion chamber is uncovered. This port releases the compressed air-fuel mixture in the crankcase so it can flow through the bypass and port opening into the combustion chamber. As the piston continues upward it covers the intake port so the air-fuel mixture in the combustion chamber is further compressed. As the piston reaches the top of the compression stroke, the mixture is ignited by a spark across the gap of the spark plug. This rapid burning of the fuel (or explosion) forces the piston downward on the power stroke.

Power is not delivered for the entire length of the stroke, because the cylinder needs some time to get rid of the remaining gases and then take on a fresh fuel charge from the crankcase for the next power stroke. As the piston nears the bottom of its stroke, the exhaust port opening is uncovered slightly ahead of the intake port opening. This movement takes advantage of the pressure within the combustion chamber, which is still comparatively high, thus allowing the remaining gases to start escaping. Further travel downward uncovers the intake bypass port, which permits the compressed fuel charge from the crankcase to enter the combustion chamber. The incoming charge assists in forcing the exhaust gases out of the cylinder to complete the cycle. The continuous repetition of this cycle produces a constant and smooth flow of power.

As you can see, the series of events in this engine operation depends largely on the location of the ports in the cylinder wall in relation to the piston's position. This type of engine is much simpler than a four-stroke cycle engine in that it does not require a camshaft and valve arrangement. When a valve, generally a reed valve, is used to open and close the port between the crankcase and the carburetor, the valve is operated by vacuum and closed by spring tension.

The crankshaft and crankshaft mounting in the crankcase are the same as in the four-stroke cycle engine, as is the ignition system.

CHARACTERISTICS OF THE TWO-STROKE CYCLE ENGINE

Two-stroke engines have certain advantages that make them desirable in specific applications where low cost and simplicity of operation are important. The engine has fewer moving parts than the four-stroke cycle engine, and it delivers a power stroke for each crankshaft revolution resulting in smooth operation.

Other considerations have an effect on operation and power output. Since the engine delivers twice as many power strokes in a given number of revolutions, compared to a four-stroke cycle engine, logically the engine would deliver twice as much power. Yet due to the problems with exhausting and filling the cylinder with a fresh fuel charge, the engine develops less power than the theoretical power advantage. These problems also result in a lower compression ratio. Due to these scavenging problems—that is, getting rid of the exhaust gases—the incoming fuel is diluted to some extent with the remaining exhaust gases. Hence further inefficiency results.

3

ENGINE POWER DEVELOPMENT AND LOSS OF POWER

HORSEPOWER DEVELOPMENT
FACTORS AFFECTING HORSEPOWER DEVELOPMENT
COMMON ENGINE PROBLEMS
ENGINE TROUBLE SHOOTING
IGNITION AND FUEL SYSTEMS
COMPRESSION
CHECKING COMPRESSION WITH A GAUGE
CHECKING COMPRESSION WITHOUT A GAUGE
CHECKING COMPRESSION ON ENGINES WITH AUTOMATIC COMPRESSION RELEASES
CRANKCASE LEAKS
INSPECTING TWO-STROKE CYCLE ENGINES FOR CARBON DEPOSITS, BROKEN RINGS, SCORED CYLINDER

For an engine to operate, three things are necessary:

1. *Compression:* The air-fuel mixture must be drawn into the combustion chamber and compressed. Enough air-fuel mixture must be compressed to meet load demands.

2. *Air-fuel:* The fuel system must provide air and fuel in the correct ratio and in the right amount to meet the load and speed conditions designed into the engine.

3. *Ignition:* A high-voltage spark, strong enough to jump across the spark plug gap against the resistance of compression, must be delivered into the combustion chamber at exactly the right time to ignite the air-fuel mixture.

All three factors must be present for an engine to operate. If any one is not performing properly, it adversely affects engine operation and efficiency. Keep this fact in mind when you are trying to locate troubles in the engine or attempting to improve or maintain efficiency.

This chapter deals primarily with basic engine functions that determine power output. Along with transmitting the force of the explosion and converting it into rotary power, the development and maintaining of compression is one of the requirements for a performing engine. In fact, it is a major function. So we will cover the factors that affect compression, along with ways of determining compression.

HORSEPOWER DEVELOPMENT

Small gasoline engines are usually rated by their horsepower, which is a unit of measurement used to indicate the amount of power the engine should be capable of developing. Horsepower is normally measured with a dynamometer under actual load conditions. Higher-horsepower ratings generally mean that the engine should be able to produce more work. In most cases, the relationships are as follows: the more horsepower, the larger the engine, the more it weighs, and the more fuel it uses. Manufacturers take these relationships into consideration when installing power plants in their equipment; they use engines with horsepower ratings that meet the necessary load ranges.

Horsepower capability is designed into the engine with certain fixed features that do not change. There are some other things over which the operator has control, such as throttle setting. Other conditions change with engine wear.

Horsepower is based on the following factors:

1. *The length of the stroke*, which is the distance the piston travels from BDC (bottom dead center) to TDC (top dead center).
2. *The bore*, which is the diameter of the cylinder.
3. *The number of strokes per minute* (rpm).
4. *The number of cylinders.*
5. *The mean effective pressure*, which is the pressure exerted on the piston when the air-fuel mixture is burned and which takes into consideration the compression ratio.

FACTORS AFFECTING HORSEPOWER DEVELOPMENT

Of primary importance to power development is the fact that wear takes place with use and that wear is speeded up with abuse. So before making the needed repairs, you must be able to determine what has happened to the engine.

The power an engine develops is otherwise affected by a number of factors, most of which are designed into the engine and do not change. Yet you should be aware of these factors when comparing features of different engines or when determining why one engine puts out more power than another. These features are as follows:

1. *Piston displacement* refers to the amount of space through which the piston travels from BDC to TDC (a stroke) and the area of the cylinder (bore). Displacement determines the amount of air-fuel mixture that can be brought into the combustion chamber, and for all practical purposes it does not change. It is expressed in cubic centimeters (cc) or in cubic inches (cu in).

2. The *compression ratio* is the ratio between the volume in the cylinder when the piston is at BDC and the volume when the piston is at TDC. The total volume is the piston displacement plus the clearance volume. The clearance volume is the space between the top of the piston when it is at TDC and the space in the cylinder head. The compression ratio is therefore the amount that the fuel charge is compressed from the original volume. If the mixture is compressed to one-eighth of its original volume, the compresion ratio would be rated as 8 to 1. Any increase in the compression ratio results in greater compression: the higher the compression, the more power that is developed (within limits). Too high compression can result in self-ignition of the fuel and/or detonation (knocking).

A number of things can affect compression: (1) a build-up of carbon in the cylinder head and on top of the piston increases compression. (2) Worn or broken piston rings can result in reduced compression through blow-by. (3) Burned and/or leaking valves result in loss of compression. (4) A leaking cylinder head gasket lets pressure escape. (5) In two-stroke cycle engines, pressure also escapes as a result of leaking crankcase seals and/or gaskets.

3. The *amount of air-fuel mixture* taken into the engine also affects output. The proportion of air to fuel is controlled by the design of the carburetor, while the amount drawn into the engine depends on the throttle opening, which is in turn controlled by the operator.

4. The *condition of the engine* also affects the amount of vacuum the engine develops, since the vacuum draws fuel into the engine.

5. *Valve timing* and *ignition timing* have a definite bearing on engine operation. Firing the fuel charge at exactly the right time relative to piston position is extremely important.

6. *Abnormal engine friction (drag)* results in power loss. Drag can result from a lack of lubrication, too much oil, a lack of clearance between moving parts such as tight bearings, or a piston with too little clearance, and so on.

7. *Incomplete scavenging* of the exhaust gases from the cylinder in a two-stroke cycle engine can also cause power loss, because the unexpelled exhaust gas takes up space needed for the incoming air-fuel mixture. It also dilutes the mixture.

COMMON ENGINE PROBLEMS

The small gasoline engine is a precision-built machine that is designed to give dependable service for a long time. But through usage, unless you make adjustments and repairs, engine parts lose their precise fit. When they do, the engine no longer runs efficiently, and problems crop up. The more common problems include a failure to start, misfiring (or running roughly), a lack of power, and running hot or abnormally noisily,

But there are problems and there are problems. Troubles that arise from any system other than the basic engine itself—such as the ignition or fuel system—can often be corrected with a little repair and/or maintenance work. (Each such system is treated in detail in later chapters.) But problems with the basic engine—the crankshaft, cylinder, crankcase, valves, piston, or any part of the engine assembly—usually call for the most time, work, and expense, not to mention special tools. Unless the basic engine parts are functioning properly, no amount of servicing on the other systems will make the engine run satisfactorily. So, although all engine troubles are definitely interrelated in one way or the other, let's deal with basic engine troubles first.

ENGINE TROUBLE SHOOTING

Some of the troubles that indicate basic engine malfunctions are:

Problem	*Cause*
Engine knocks	Excessive carbon in combustion chamber
	Loose or worn connecting rod bearings
	Loose flywheel
	Worn cylinder bore or piston
	Improper ignition timing
Engine fails to start or runs with difficulty	Poor compression
Engine misses under load	Improper valve clearance
	Weak valve springs
	Poor compression

Problem	Cause
Engine lacks power	Worn piston rings
	Valves leaking
	Poor compression
Engine overheats	Cooling fins clogged
	Air flow obstructed
	Excessive carbon in engine
	Lack of lubrication
Engine vibrates excessively	Bent crankshaft
Engine uses excessive oil	Leakage—gaskets and/or seals
	Piston rings and/or grooves worn
	Valve guides worn
	Cylinder walls worn, glazed, or scored
	Oil control ring return holes clogged
	Poor compression

For an engine to operate efficiently, certain units must have gas-tight seals, while others must be oil-tight. Moving parts need a certain amount of clearance—that is, space between moving parts—to provide for lubrication and expansion created by heat. Generally both lateral (side) and end clearances must be taken into consideration. On the other hand, excessive clearance is a direct result of engine wear. Too much clearance can cause a number of problems, such as a loss of power, hard starting, excessive oil consumption, or inefficient, as well as noisy, operation. In addition, excessive clearance can bring about knocking and other noises that may in turn damage additional parts. Engine disassembly and repair are necessary to restore the proper clearances and the original efficiency. Parts may have to be replaced rather than repaired if not properly serviced.

The repair may involve nothing more than tightening a bolt, or it could mean the complete disassembly down to the last unit. Since you don't ordinarily know the extent of work needed at first, before attempting any type of repair, always locate the problem. Use a systematic approach to isolate the problem and its cause. Start with units on the outside of the engine, because they are easy to get at, before disassembling any of the more complex units. Through a process of elimination, starting with the elementary components that are easily accessible, you can determine which are functioning properly. Work from the known to the unknown, from the simple to the more complex.

Consider also that some problems can be caused just as well by malfunctions in the ignition and/or fuel system as by those in the engine itself: lack of power, hard starting, rough running, or poor idle. Yet an oil leak would obviously be a basic engine problem, and so would excessive exhaust smoke that is gray, not black, accompanied by excessive oil consumption and knocking. Pre-ignition and detonation can also be the result of carbon build-up within the cylinder.

If certain noises occur while the engine is running, they may indicate wear or malfunctions within the engine. Sometimes describing or isolating the noise is hard, but you should be able to get an idea of what is making the noise by listening carefully to the engine under different operating conditions.

1. A *ping* that occurs under load conditions may be caused by pre-ignition or detonation resulting from carbon build-up in the combustion chamber. A "ping" when the engine first starts up could be the result of timing that is advanced too much.

2. A *rap*, like a hammer striking a hard surface when the engine is loaded and then released, is caused by a loose connecting rod bearing.

3. A steady *click* may be the result of excessive clearance between the tappet and the end of the valve stem.

4. A *thump* or thud when the engine is under load could indicate worn crankshaft bearings.

5. A *slap*, not as pronounced as a rap, could be the result of a piston with too much clearance. It is generally more noticeable when first starting up.

Even if you conclude that the problem is within the basic engine, always try to determine the component causing the trouble before disassembling the engine.

IGNITION AND FUEL SYSTEMS

Although malfunctions within the engine can prevent starting, the ignition or fuel system is generally at the root of the problem when an engine fails to start. Unless you are positive that the engine is at fault, make a quick check of the ignition and fuel system.

To check the *ignition system*, disconnect the spark plug wire from the spark plug. With the ignition switch on, place the end of the spark plug terminal about 3/16 to 1/4 inch from a good ground. Usually a clean spot on the cylinder head is a good ground. Figure 3-1 illustrates how to check for a spark to the spark plug. Be sure to hold the insulated part of the wire, *not* the terminal, or you may get a shock when the engine is cranked. When you crank the engine, a spark should jump the gap between the terminal and ground. If you get no spark, the trouble is in the ignition system. Removing the spark plug permits you to turn the engine much easier and faster. If a good spark occurs, chances are the ignition system is functioning correctly.

To check the *fuel system*, make sure that there is gasoline in the tank and that the shut-off valve is open. Close the choke tightly. Crank the engine for several revolutions. Remove the spark plug and inspect it. If the plug is wet, fuel is getting into the combustion chamber. If the spark plug is dry, pour a small amount (one-half teaspoon) of gasoline through the spark plug hole into the engine. Replace the spark plug. Crank the engine with the switch on and the choke open. If the engine fires a few times and then quits, the problem is in the fuel system.

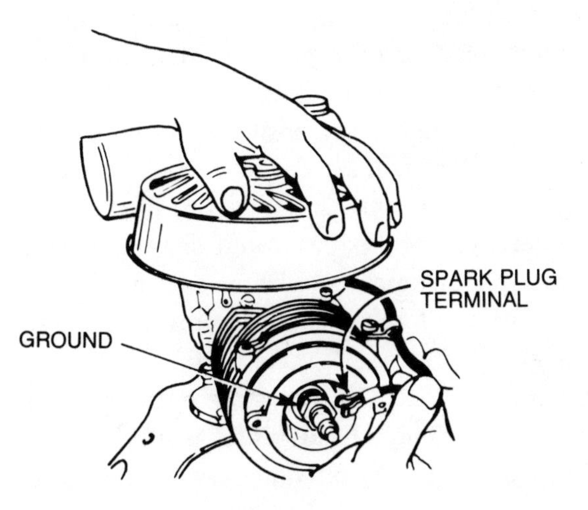

FIGURE 3-1
Checking for spark
Courtesy Tecumseh Products Co.

COMPRESSION

If, when cranking the engine, you feel no resistance or impulses, the engine could be lacking compression. So always check the compression before disassembling an engine or whenever diagnosing engine troubles. A compression check determines the compression pressure within the cylinder. Compression is determined by how well the piston rings seal, whether or not the valves are properly seating, and whether or not the cylinder head gasket is tight. In the case of the two-stroke cycle engine, leaking crankcase gaskets or oil seals result in the loss of compression.

Low compression results in a rough running engine (particularly at low speeds), an engine that is hard to start, power loss, and excessive smoking.

In a one-cylinder engine compression is generally thought of in terms of power output rather than in terms of pounds per square inch. Yet some manufacturers recommend that you make a compression test and check it against specifications. Obtaining an accurate compression reading with a compression gauge may be difficult, because many machines do not have a starting motor to crank the engine at a constant speed and because the displacement of the engine is so small. Regardless of its accuracy, a reading taken with a compression gauge gives you at least some idea of the amount of pressure present.

CHECKING COMPRESSION WITH A GAUGE

You need a compression gauge to get a compression pressure reading. A gauge can be purchased for a reasonable price at a store that sells automotive parts and accessories. Maybe you can borrow a tester or have someone at a garage make the test. The operation itself is very simple.

If possible, the engine should be warm. Shut off the ignition switch. Disconnect the spark plug wire. Clean the spark plug recess in the cylinder head with compressed air or wipe out with a rag so dirt does not fall into the engine when you remove the plug. The throttle valve and choke valve should be wide open. Insert the compression gauge into the spark plug hole (Figure 3-2). Crank the engine as fast as possible for approximately 8 to 10 revolutions. Record the reading. If the reading is low, remove the gauge. Squirt about one tablespoon of engine oil on top of the piston, making sure no oil gets on the valves. Recheck the compression. If the reading improves considerably, it indicates that the piston rings are leaking. If the reading improves only a very small amount, then the valves are either sticking or not sealing properly. If the cylinder head gasket is leaking, you can usually see oil or hear leakage to the outside between the cylinder barrel and cylinder head.

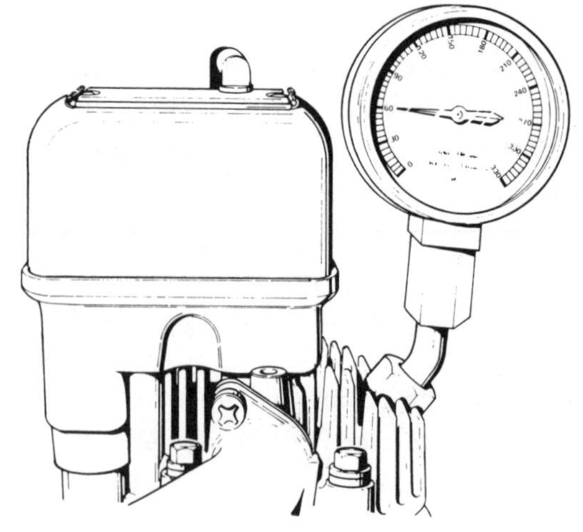

FIGURE 3-2
Checking compression with a compression gauge
Courtesy Tecumseh Products Co.

The compression reading for a four-stroke cycle engine should be approximately 100 to 150 psi, and for a two-stroke cycle engine approximately 90 to 100 psi. A reading much below 75 psi indicates trouble.

The two-stroke cycle engine does not have valves in the top of the cylinder block or cylinder head to control mixture and exhaust flow. But some engines of this type have a decompression valve located in the cylinder head. The valve and spring, along with a control button or cable, can be easily identified on the cylinder head. After a while the valve may fail to seat tightly, resulting in a compression leak. To check the valve for leakage, cover the area around the valve stem (where leakage might occur) with soap suds. Crank the engine to build up compression. Be sure the valve is closed. If bubbles occur, the valve is leaking.

CHECKING COMPRESSION WITHOUT A GAUGE

You can get an idea of the compression pressure without a gauge. With the spark plug removed, be sure that the choke valve and throttle valve are open. Then place your thumb over the spark plug hole. You should feel air pressure as the engine is cranked: the more pressure against your thumb, the greater the compression. Squirting oil on the top of the piston should increase the pressure if the piston and/or rings are not sealing properly.

Another quick test to determine compression, with the spark plug installed, can be made by turning the flywheel slowly until you feel pressure. When you feel pressure, give the flywheel a quick spin. If the flywheel snaps back strongly, due to resistance, then you have compression—at least enough to start the engine. If the flywheel keeps turning or stops weakly without snapping back, the compression is low.

CHECKING COMPRESSION ON ENGINES WITH AUTOMATIC COMPRESSION RELEASES

The compression on engines with automatic compression releases cannot be checked in the regular manner. To check compression by feel, rotate the engine backward against the power stroke. You should feel resistance while it's turning. If compression is low or nonexistent, then you will feel no resistance.

If you want to check the compression using a gauge, the engine must also be rotated backward.

CRANKCASE LEAKS

The crankcase of the two-stroke cycle engine must be air-tight, because pressure must be developed in the crankcase to force the air-fuel mixture into the combustion chamber when the transfer port is uncovered. Leakage affects the amount of air-fuel mixture getting into the combustion chamber. Too little fuel results in a starved engine and a power loss. To determine if there is a leakage, look for evidence of oil and fuel at the crankcase parting surfaces, at the crankcase seals at the crankshaft, and at other joints. If you brush soap suds on the various joints while the engine is running, bubbles may appear at the leakage point.

You can make a more complete pressure test to locate leaks in the following manner. Place a rubber or plastic sheet between the exhaust manifold and cylinder to form an air-tight seal. Do the same between the intake manifold and the cylinder. Connect a pressure regulator to the impulse fitting or another opening into the crankcase. Apply compressed air until the pressure gauge on the regulator reads 7 psi. Then shut off the air. The pressure should not drop below 5 psi for at least 10 seconds. If the pressure drops rapidly, maintain 7 psi pressure on the crankcase while applying a soap suds solution to the seals and seams until you find the leak. Figure 3-3 shows a pressure tester being used to check a two-stroke cycle engine crankshaft seals and gaskets for leaks.

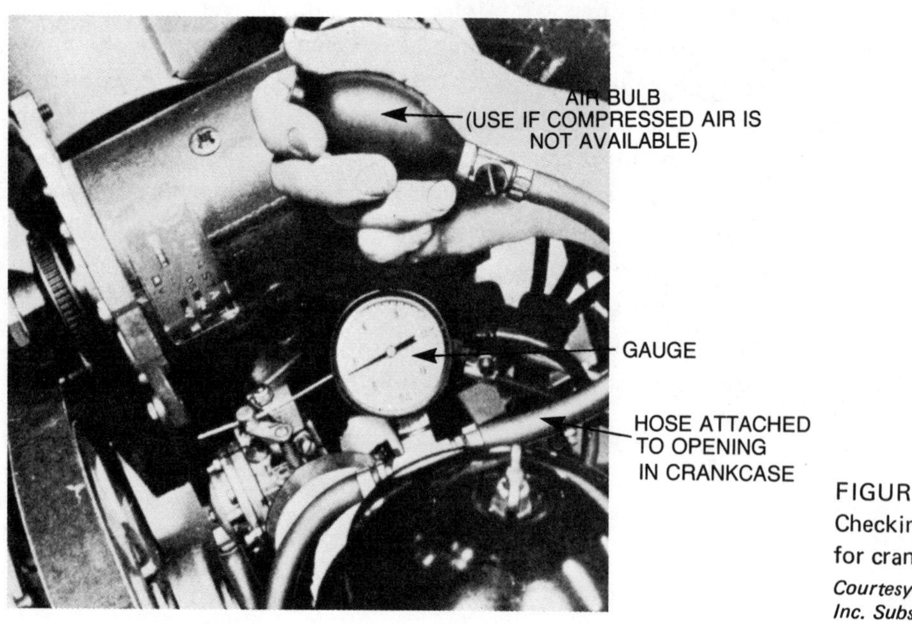

FIGURE 3-3
Checking two-stroke cycle engine for crankcase leaks
Courtesy Harley-Davidson Motor Co., Inc. Subsidary of AMF Inc.

INSPECTING TWO-STROKE CYCLE ENGINES FOR CARBON DEPOSITS, BROKEN RINGS, SCORED CYLINDERS

If you suspect internal trouble in a two-stroke cycle engine, remove the exhaust manifold assembly before attempting a major disassembly operation. This simple step permits you to determine the amount of carbon in the exhaust outlet as well as on top of the piston and in the combustion chamber. Some carbon is normal, but thick hard carbon deposits on the top of the piston and combustion chamber can mean the loss of power and overheating. Excessive carbon build-up can also result in raising the compression ratio, which can in turn cause detonation (knocking) and damage to the piston. You can remove most of the carbon through the exhaust opening by first scraping it loose with a knife blade or narrow putty knife and then blowing it out with air. Do not mar or groove the metal parts of the cylinder head or piston top.

With the exhaust manifold removed, you can examine part of the piston rings and part of the piston as well as some of the cylinder wall. The piston rings should be uniformly bright. When the rings are gently pushed back into the grooves, they should spring back. The cylinder wall should be smooth and free from scores. The piston should also be smooth and free of scores, and the ring lands should not be damaged in any way.

BASIC ENGINE SERVICE

ENGINE WEAR
ENGINE DISASSEMBLY

The next three chapters include descriptions of the various engine parts, along with explanations of how they are disassembled, inspected, repaired, and put back together. This information covers the basic engine parts common to all small gasoline engines of both the two-stroke and the four-stroke cycle types.

Some of the operations you may not want to do, and others require special equipment. But a number of operations you should be able to do quite readily with the necessary know-how. Although instructions for the *complete* disassembly and inspection of an engine are covered, in many cases you need to disassemble only to the extent that the unit causing the problem can be serviced.

The repair operations that require special equipment are usually better left to the professional. These include such tasks as honing the cylinder barrel, refacing valves, grinding valve seats, regrinding the crankshaft, and the like. When you determine, for example, that the valves need reconditioning or that the cylinder must be honed to remove scores or scratches, you can usually make considerable savings just by taking the parts to be serviced at a shop. (Most automotive machine shops can remachine small engine parts.) Doing so saves the labor charge for disassembly and assembly.

One last tip: So you can work easily on the engine, remove it from the machine. Uncoupling it from the driven unit is usually a matter of removing the bolts in the base that hold the engine to the frame. The chain, belt, or drive coupling can usually be removed or disconnected after the engine is loosened.

ENGINE WEAR

Although engines are precision-built machines designed to give long and dependable service, many of their moving parts begin to lose their precision fit after a period of operation. Engine parts are made of metal to provide strength and to conduct heat. But because of the high temperatures that the engine develops, the metal parts expand. The expansion and the contraction, plus heavy loading and varying speeds, cause the parts to lose their original clearances. This "looseness" can result in loss of power, starting problems, and/or abnormal noise.

Some engine parts are fitted to provide a gas-tight seal between the piston and cylinder wall and between the valve face and seat. Excessive clearance between the piston and cylinder is a result of wear. Improper seating of the valves is caused by heat and wear, resulting in valve leakage and loss of compression. Low compression results in a number of problems, such as poor performance, excessive oil consumption, hard starting, and the like. With any of these conditions, the engine should be serviced so as to restore efficient operation.

Keep in mind also that what might appear to be an engine performance problem could actually be the fault of the equipment driven by the engine rather than the engine itself. Various types of equipment are powered by small gasoline engines, and so many different problems can occur. Some of the more common equipment problems are as follows:

1. Hard starting, kickback, or the failure to start may be the result of a loose mower blade or a loose belt, whose backlash counteracts the engines cranking effort.

2. Starting "under load" (that is, when the engine is hard to turn over) may occur because the equipment is not totally disengaging from the engine.

3. A vibration may be caused by a bent or out-of-balance blade or by a worn coupling that allows a blade to shift. Loose mounting bolts can also result in vibrations, as can a cracked mounting plate or deck.

4. Any binding or drag in a driven unit can affect engine performance. To check out this possibility, completely disengage the driven unit to see if this makes a difference in engine operation.

5. A noise can be the result of loose driven blades, pulleys, sprockets, shafts, or something else. The lack of lubrication might be the cause.

Of major importance is that you locate the trouble *before* disassembling the engine. This way you eliminate unnecessary work, and you know what to look for as you take the engine apart.

ENGINE DISASSEMBLY

Before disassembling an engine, you should have the necessary gaskets available. In most cases gaskets are ruined when you take the engine apart. You should replace the cylinder head gasket any time the head is removed. Generally the used gasket has lost its elasticity and no longer seals properly. Most important is that all surfaces requiring gaskets do not leak after reassembly.

An owner/operator's manual is helpful, because it usually provides specifications and information on required maintenance and adjustments, such as spark plug settings and carburetor adjustments.

During the disassembly process, check and service other units that have to be removed. If the gas tank must be removed, always check and clean the filter and/or screen if necessary. Whenever the spark plug is removed, clean and regap it. Do the same for the ignition contact points, air cleaner, and carburetor screen. The few extra minutes it takes, you are assured a good operating engine.

Cleanliness is essential. Since you are dealing with precision-fit parts, any type of abrasive (grit, dirt, and the like) that gets into the engine only hastens wear and may plug small openings. For this reason clean the outside of the engine so as to eliminate the possibility of dirt getting into the engine or onto parts as you remove them. A number of commercial cleaning solutions are available that dissolve grease and oil. In most cases the solvent can be brushed or sprayed on and flushed off with water. Some household detergents will also loosen dirt and grease.

The order of disassembly may vary somewhat from engine to engine. If you remove the parts in some sort of logical sequence, you will make it easier to remember where the parts go when reassembling them. Using different containers and isolating certain sections of the engine as it is disassembled may also be helpful. This technique is of assistance in remembering where the different parts belong.

Despite the variations among the different engines, they all have many of the same components. The sizes, shapes, and locations of some parts may be different, but in general the service operations (disassembly, inspection, and disassembly) remain much the same.

What you determine needs to be done will govern the extent of disassembly. If the lower section of the engine—containing the crankshaft, connecting rod, and those parts—is to be serviced or if the engine is to be tipped, drain the oil from the engine (four-stroke cycle only). To do so, remove the oil drain plug at the bottom of the crankcase, and let all the oil drain completely before replacing the plug.

During disassembly of the engine, carefully observe the condition of each unit as it is removed and/or disassembled. Figure 4-1 shows the disassembled engine. Many times certain general conditions or the appearances of a particular part indicate that the engine has been used hard and/or that it has not been properly maintained.

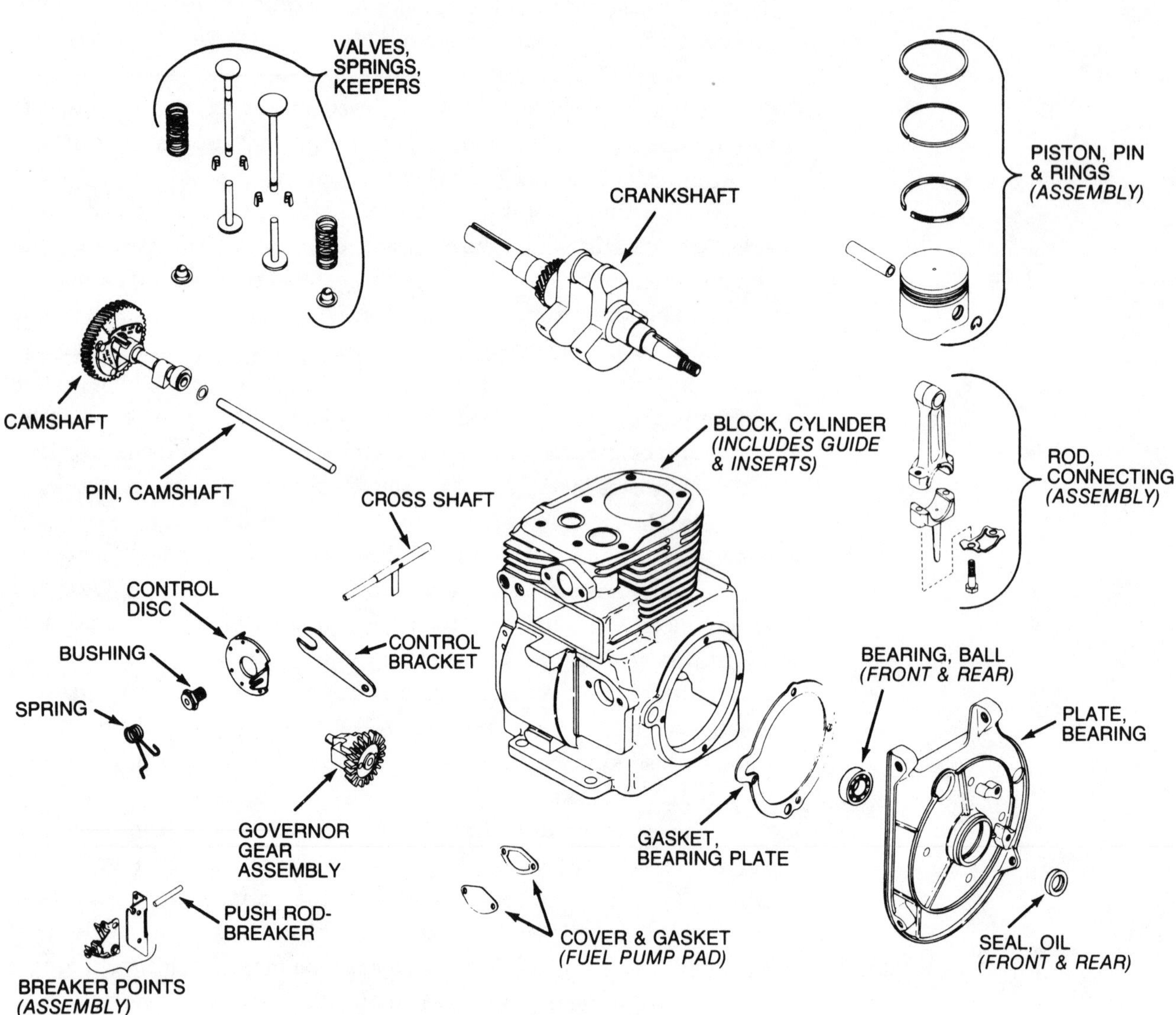

FIGURE 4-1
Disassembled four-stroke cycle engine
Courtesy Kohler Co.

In such cases, improved operating conditions and better maintenance could have prolonged the service life of the engine. Some of the conditions to look for include excessive sludge in the valve spring housing, as well in the crankcase. If you see a scored cylinder wall, piston damage, and signs of external oil leakage, you should find out why these things happened.

Excessive sludge—some sludge is normal—could indicate several things. The most common cause is to infrequent oil changes. It could also be that the engine has been operating with an improperly adjusted ignition timing. An overly rich fuel mixture, caused by an improper carburetor adjustment, can also result in sludge formation. The same thing can happen if an air cleaner or crankcase breather is clogged.

Unburned fuel (too rich a mixture) or over-choking not only adds to sludge formation, but it can also cause scuffing and scoring (grooving) of the cylinder and/or piston. As raw fuel seeps down the cylinder wall it washes away the lubrication from the cylinder as well as from the piston. The piston rings then make metal-to-metal contact, which can result in scoring and scuffing. If the piston or piston rings were installed without enough clearance for expansion, scuffing could take place when the engine gets hot. Localized hot spots may occur in the cylinder wall because of clogged cooling fins or inadequate lubrication, and these hot spots can cause scoring. A broken ring can also make grooves in the cylinder wall.

Piston damage can take various forms. The top ring may be burned through, including the piston ring land (the metal between ring grooves). The rings may be stuck in their grooves, or some of the rings may be broken. This condition can be the result of abnormal combustion, but generally it is caused by advancing the ignition timing too much. In turn, the condition may result in preignition, which raises the temperature to the point that it can burn the metal in the piston and/or rings.

If excessive oil leakage is evident, it could be the result of a clogged breather, assuming all the bolts are tight in the crankcase. Normally all four-stroke cycle engines operate with a slight vacuum within the crankcase, which is brought about by the moving parts inside the engine. If the crankcase breather (generally part of the valve tappet cover) is clogged, or if the valve in the breather is stuck, pressure may build up in the crankcase. The same thing can happen if there is excessive blow-by, that is, if part of the explosive force is escaping past the piston and rings. Blow-by tends to force oil out through oil seals, gaskets, or other openings in the crankcase.

Discovering and eliminating the cause of the problem or malfunction should help to prevent the same problem from happening again.

REMOVE AIR CLEANER
REMOVE CARBURETOR
REMOVE EXHAUST SYSTEM
REMOVE BLOWER HOUSING
REMOVE CYLINDER HEAD
REMOVE VALVE COVER (BREATHER)
REMOVE AND RECONDITION VALVES
SERVICE DECOMPRESSION VALVE LOCATED IN CYLINDER HEAD

If the engine is to be disassembled for service, then in most cases you have to remove the air cleaner, the fuel tank, carburetor, exhaust system, blower housing, and cylinder head. If the valves leak, they need to be removed from the cylinder block or cylinder head for service. Even without valve problems (two-stroke cycle engines do not use valves to control fuel and exhaust flow), you have to remove the cylinder head and other components to service the rest of the engine.

REMOVE AIR CLEANER

Remove the air cleaner. The cover is generally held in place by a wing nut or screw, or it is merely snapped on and can easily be pried off. (Figure 5-1 shows an air cleaner that has a snap-type cover.) If necessary, remove any remaining screws that attach the cleaner to the engine. Always clean the air filter (cleaner) before reinstalling it. This procedure is discussed in detail in Chapter 8.

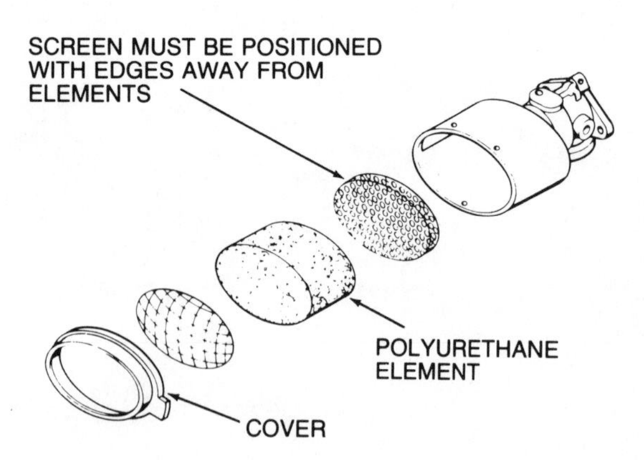

FIGURE 5-1
Snap cover air cleaner with polyurethane element
Courtesy Tecumseh Products Co.

REMOVE CARBURETOR

Disconnect the fuel lines and drain the fuel from the tank. Remember that gasoline and gasoline fumes are highly inflammable. Keep all open flames and sparks away from the gasoline.

If necessary, remove any shrouding or control panels to provide access to the carburetor. Look for signs either that air is leaking or that mounting gaskets have loosened, deteriorated, or otherwise been damaged. If so, repair the carburetor before reinstalling it. Very carefully note the location of the governor spring, linkage, throttle control, and choke control. Do not bend the linkage or stretch the springs. The linkage and spring must be replaced in exactly the same position from which it was removed. Figure 5-2 is a typical throttle and governor linkage arrangement. Mark the linkage so you know where it goes when you are reinstalling the carburetor. Do not bend or distort the linkage. Remove the carburetor. Figure 5-3 shows the air cleaner, exhaust outlet, and carburetor attached to the engine. The carburetor linkage is also shown.

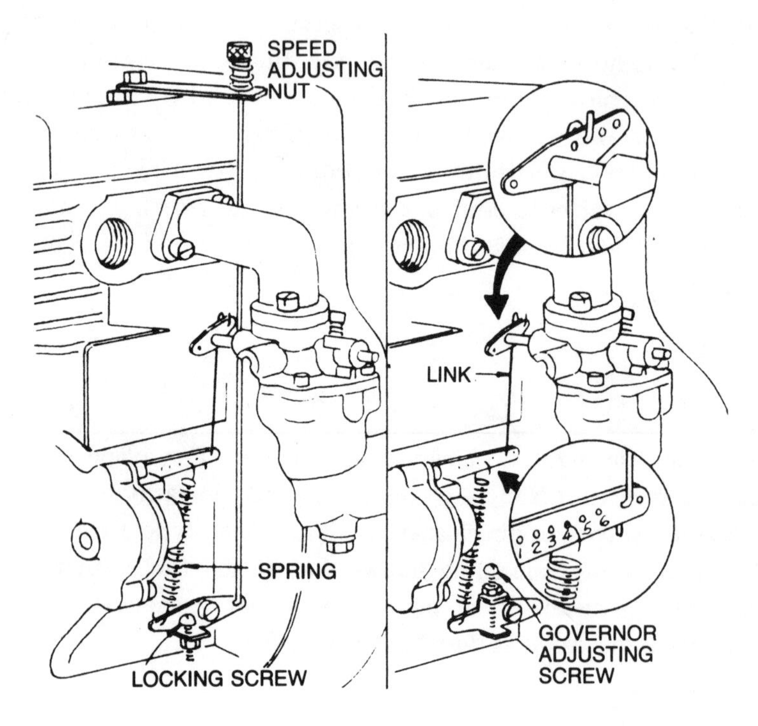

FIGURE 5-2
Governor throttle controls
Courtesy Briggs & Stratton Corp.

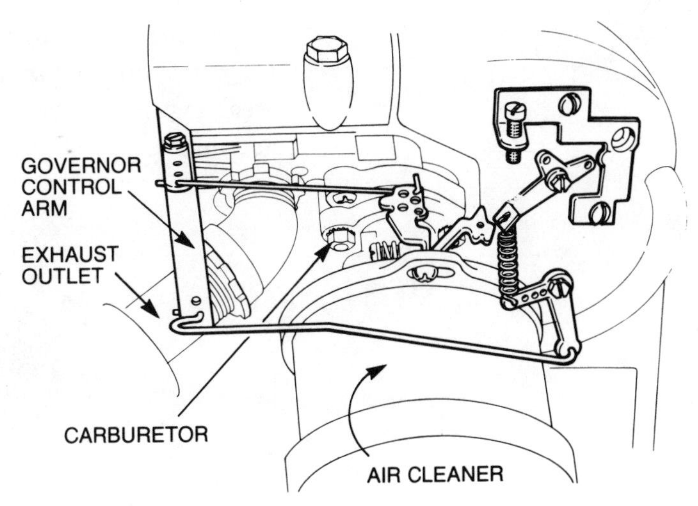

FIGURE 5-3
Carburetor and related components
Courtesy Tecumseh Products Co.

When disassembling a two-stroke cycle engine that has a reed valve, remove the valve after removing the carburetor. Note how the valve is positioned on the engine so it can be reinstalled in the same manner. This one-way valve must be installed in the proper direction in order to function. The engine equipped with a reed valve has the carburetor mounted directly on the crankcase. The two-stroke cycle ported engine has the carburetor mounted on an intake manifold.

REMOVE EXHAUST SYSTEM

Remove the exhaust manifold and/or muffler. The unit may be attached to the cylinder with bolts, or it may screw into the cylinder. On an engine with overhead valves, the exhaust outlet is attached to the cylinder head.

REMOVE BLOWER HOUSING

Remove the blower housing. Generally bolted to the cylinder head and engine block, the blower housing includes the rope starter assembly or impulse starter, whichever type is used. Exercise care when removing the housing so the starter spring does not release. If an electric starter or combination starter/generator is used, disconnect the battery cables at the battery terminals before disconnecting the wires from the starter/generator unit. Then remove the blower housing.

REMOVER CYLINDER HEAD

Remove the bolts that attach the cylinder head to the cylinder barrel. Note difference in the length of the bolts, if any. If so, make sure the bolts are replaced in the same holes from which they were removed. Remove the cylinder head. Carefully clean the carbon deposits from the cylinder head, the top of the piston, and the valves.

Use a putty knife or similar blade to scrape off the deposits. Be careful not to nick or scratch the valve faces or seats or the gasket seat area. Figure 5-4 shows a cylinder head, cylinder block, carburetor, and muffler.

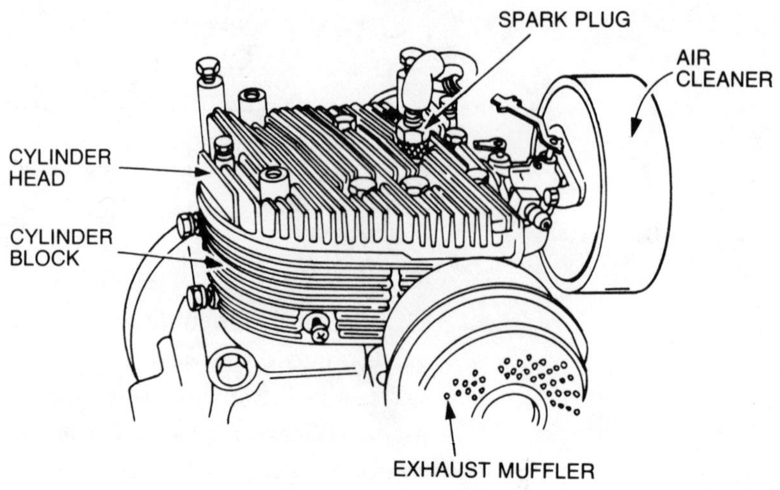

FIGURE 5-4
Cylinder head mounted on cylinder block
Courtesy Tecumseh Products Co.

The valves on a few engines are located in the cylinder head. Push rods, located between the camshaft and rocker arms, open the valves. Figure 5-5 shows the rocker arm assembly located on the cylinder head; the rocker arms actuate the valves. The valve rocker arm assembly must be removed from the cylinder head before you remove the head and service the valves.

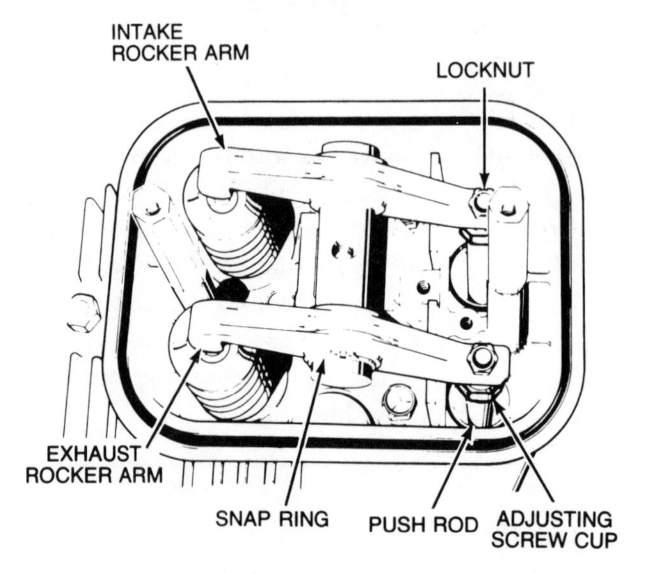

FIGURE 5-5
Rocker arm assembly on overhead valve engine
Courtesy Tecumseh Products Co.

The two-stroke cycle engine does not have poppet valves on the top of the cylinder block, but its exhaust port must be carefully cleaned.

REMOVE VALVE COVER (BREATHER)

Other than removing the valve cover, no further disassembly of the engine is required if only the valves are to be reconditioned. The valve cover is usually a plate that is slotted or that has an opening for a tube, to allow the crankcase to breathe. It is attached with screws or a bolt. Many times the sludge that accumulates in the chamber must be removed before you remove the valves. Do not let the sludge or cleaning agent get into the crankcase unless the engine is to be completely disassembled.

A clogged crankcase breather can make pressure build up in the crankcase. In this case, oil is then forced past crankshaft seals or up into the combustion chamber past the piston rings. The result is excessive oil consumption or leakage.

Some engines have a fiber disc valve or reed valve in the breather inlet to maintain a vacuum in the crankcase. Air can flow out of the crankcase, but the valve blocks the return flow. The valve must move freely and seat tightly. Do not force this valve to move as it can easily be distorted. Figure 5-6 shows a breather that uses a one-way valve and a tube to carry the vapor away. Many breathers are vented through the air cleaner.

Engines in good condition (four-stroke only cycle) should have a crankcase vacuum of ½ to 1 inch hg, as calibrated on a mercury vacuum gauge. Some manufacturers recommend making a vacuum test. To make a vacuum test, use a rubber stopper that fits into the oil fill hole. Drill a hole in the center of the stopper so it accepts a pipe adapter of the size that fits the vacuum gauge hose. Take a vacuum reading with the engine operating at normal temperature. This will determined if the valve is functioning properly.

FIGURE 5-6
Crankcase breather with one-way valve
Courtesy Tecumseh Products Co.

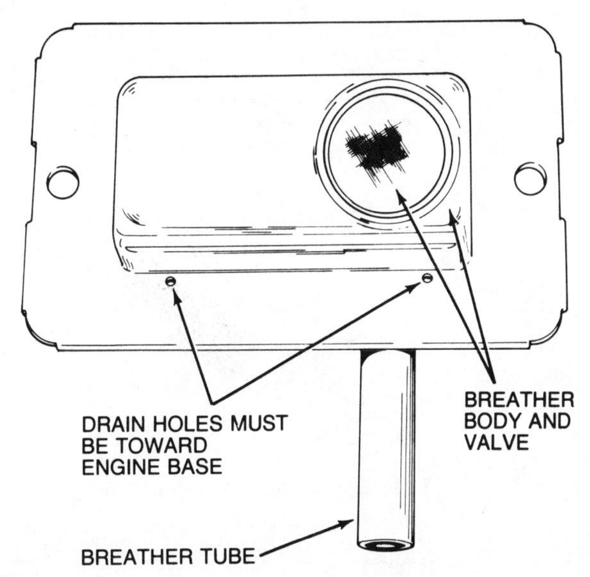

REMOVE AND RECONDITION VALVES

At least three different methods are used to hold the valve spring retainer and valve in place. These include a pin, a split C or slotted collar, or a slotted retainer. Figure 5-7 is a sectional view of a complete valve assembly showing the valve guide, valve, valve spring, spring retainer, valve seat, tappet, and valve keeper. Regardless of the method used to hold the spring and valve in place, the spring must be compressed with the valve seated in order to remove the valve spring retainer. Figure 5-8 shows the compressed valve spring, along with the collar-type keeper. A valve spring compressor can be purchased at an auto parts store.

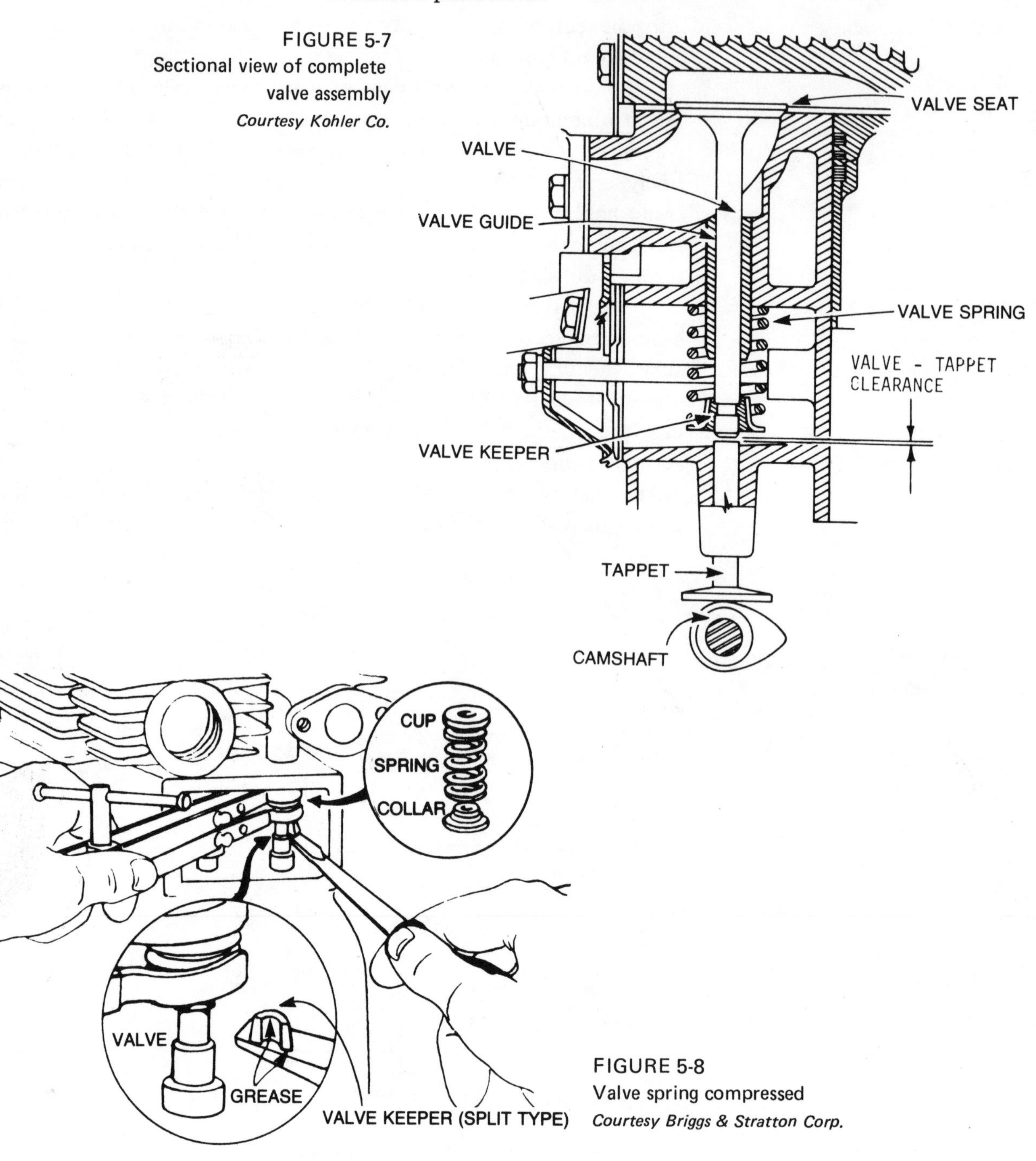

FIGURE 5-7
Sectional view of complete valve assembly
Courtesy Kohler Co.

FIGURE 5-8
Valve spring compressed
Courtesy Briggs & Stratton Corp.

Sometimes you can compress the spring and raise the retainer with a couple of screwdrivers to pry the spring up. But someone else must hold the valve down and remove the pin or collars. In the case of the slotted retainer, compress the spring above the retainer because the retainer must be moved over into the larger opening to slip the retainer over the end of the valve. The process is reversed when installing the retainer pin, collars, or slotted retainer.

The intake valve may be made of a different material from that of the exhaust valve. Always return the valve to the port from which it was removed.

Inspect the valves and valve seats after removing the carbon. You may use a putty knife or knife blade and wire brush to remove carbon.

If the valve face is warped, or if portions are cracked, burned away, or deeply pitted, the valve should be replaced. Remember the purpose of valves is to maintain a gas-tight seal. If they are not badly pitted or burned, they can be refaced to obtain a new surface that will be gas-tight. The seats must also be resurfaced.

The valve face and seat can be lapped with valve lapping compound to remove the glaze and shallow pits, as well as to clean up the surface. Too much lapping grooves the valve face and makes the seat too wide. To lap a valve, place a small amount of fine valve lapping (grinding) compound on the face of the valve. Insert the valve so its face rests on the valve seat. Then turn the valve in both directions a number of revolutions. Remove the valve and check to see if the compound is cutting all the way around. Repeat this process until you obtain a uniform seat and face. Remove all traces of compound, and then oil the valve face and seat.

If you install new valves, reface the valve seat in the block to the correct angle. When refacing the valve face, if the margin is ground to less than 1/32 inch, the valve should be replaced. Figure 5-9 shows the valve face dimensions.

The reconditioning of valves, faces, and seats requires special equipment. Most garages and all automotive machine shops have the equipment for remachining valves and valve seats. With the valves

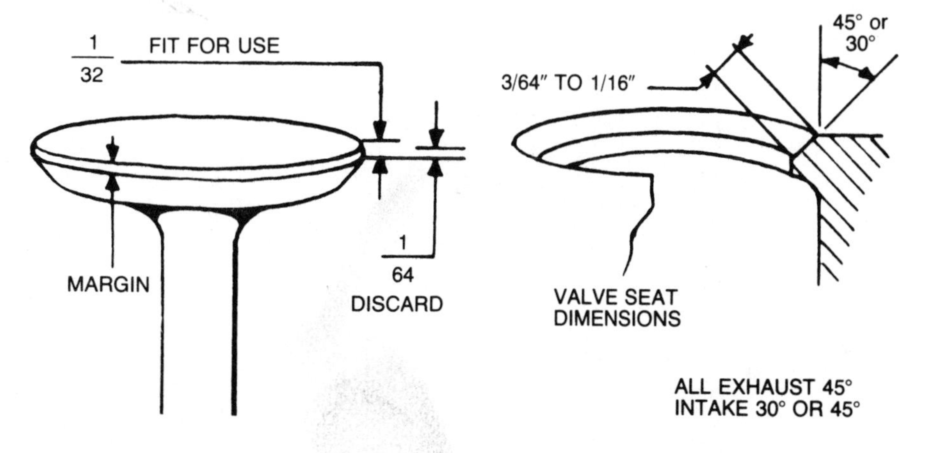

FIGURE 5-9
Valve face dimensions
Courtesy Briggs & Stratton Corp.

removed and cleaned, take them and the cylinder block to an automotive machine shop for reconditioning. This should not be too expensive.

Some engines use a valve seat insert in the block. Since the exhaust valve is subjected to considerable heat, special alloy steel valve seat inserts may be used to prolong the valve seat life. Replacing the insert is possible if it becomes badly burned or cracked, but doing so requires special equipment.

The valve springs should be removed and checked. Some engines use a different spring for the exhaust valve. Always make sure these springs are reinstalled in the same place from which they were removed. Compare one spring with the other for free height. If there is a difference, replace the springs. Also replace them if they are pitted or cracked, or if one is noticeably weaker than the other.

The valve stem must fit properly in the guide to maintain a good seal between the valve face and seat. The clearance between the intake guide and valve stem should not exceed 0.0045 inch. By moving the valve head sidewise when it is slightly raised off the seat, you can get an idea of the clearance. A dial indicator can be used to obtain an accurate reading. Most manufacturers make valves available with oversized stems. The guide can be reamed to accommodate a valve with a larger stem.

Before reinstalling the valves, also check the tappet clearance, which is the clearance between the end of the valve stem and the tappet. The valve must be tightly closed, and the lobe on the camshaft must be turned to the closed position. Check the exhaust valve when the intake valve is wide open. Then check the intake valve when the exhaust valve is wide open. A wide-open position is obtained by turning the crankshaft. Figure 5-10 shows how to check the valve clearance with a feeler gauge.

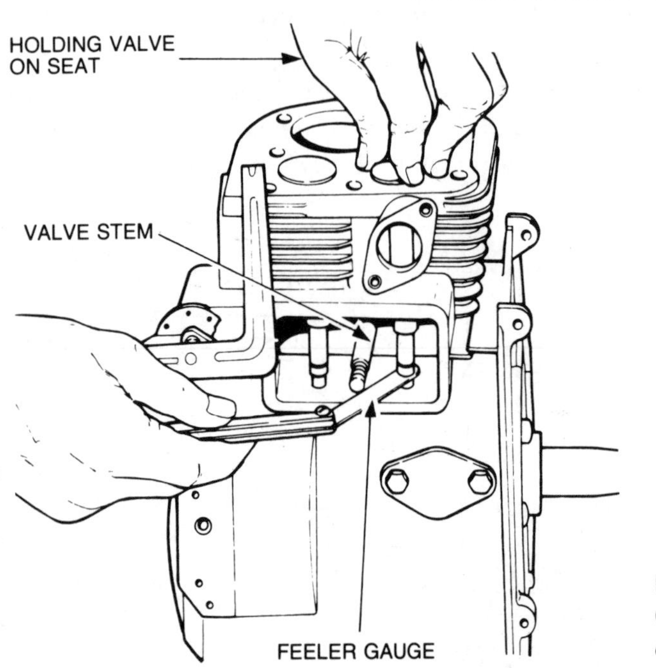

FIGURE 5-10
Checking valve clearance
Courtesy Kohler Co.

Grind the end of the valve stem to obtain the correct clearances, which is approximately 0.008 inch for the intake valve and 0.015 inch for the exhaust valve. For exact clearances always check the manufacturer's manual for the specific engine. If the engine uses an adjustable tappet, you can obtain the proper clearance by turning the adjustment up or down. If too much clearance exists, the valve should be replaced. Always adhere to the minimum clearance to prevent leakage by the valve's expanding and remaining open when the engine is hot. Excessive clearance, within reason (an additional 0.010 inch), will probably result in a noisy valve, but it will not have a noticeable affect on operation.

SERVICE DECOMPRESSION VALVE LOCATED IN CYLINDER HEAD

When a poppet-type decompression valve, located in the cylinder head of a two-stroke cycle engine, is suspected of leaking, check it. To test for leakage around the valve, install the spark plug in the cylinder head. With the combustion chamber facing upward, pour gasoline into the chamber. If the valve is not seating properly, gasoline will leak out. If there is a leak, tap the valve head lightly to see if the tapping will properly seat the valve. If it still leaks, remove and service it.

Clean the carbon off the valve face, valve seat, and valve stem. This cleaning may allow the valve to close tightly. If it does not correct the trouble, the valve face and seat can be lapped to obtain a better seal. Place valve lapping compound on the valve face. Turn the valve several revolutions with the face contacting the seat. Then clean off all traces of the compound, and recheck for leakage. Repeat the operation until the valve no longer leaks.

6

SERVICING CYLINDER, PISTON, CRANKSHAFT, AND CAMSHAFT

REMOVE FLYWHEEL
DISASSEMBLE CRANKCASE
INSPECT INTERNAL ENGINE PARTS
SERVICE INTERNAL ENGINE PARTS
INSPECT CYLINDER BORE
CHECK CONNECTING ROD BEARING CLEARANCE
MEASURE CYLINDER BORE
RECONDITION CYLINDER
CHECK PISTON
CHECK RINGS
CHECK PISTON PIN
INSTALL RINGS
INSPECT CRANKSHAFT AND CRANKSHAFT BEARINGS
INSPECT CRANKSHAFT BALANCE SYSTEM
INSPECT CAMSHAFT AND CAMSHAFT GEAR
INPSECT AUTOMATIC COMPRESSION RELEASE
INPSECT MECHANICAL GOVERNOR
INSPECT OIL PUMP DRIVE GEAR
INSTALL OIL SEALS
ASSEMBLE ENGINE

If you establish that further disassembly is necessary either to correct a malfunction or to determine the exact nature of abnormal operation, carefully examine each component as it is removed. Pay particular attention to how it is removed so you can replace it in exactly the correct position. You may want to mark some of the parts for alignment purposes. To do so, either file a small notch in corresponding parts, or use a center or prick punch to mark the parts.

After the exhaust system has been removed from a two-stroke cycle engine, you can examine part of the piston, combustion chamber, piston rings, and cylinder wall. The cylinder bore on any engine can be examined after removing the cylinder head and turning the flywheel until the piston is at BDC. Check the wall of the cylinder for rough spots, scratches, scores (grooves), and wear. Excessive wear, if any, is generally noticeable by a ridge at the top of the cylinder where the piston rings do not contact the cylinder wall.

Due to differences in construction among engines, the disassembly procedure may differ somewhat from one engine to another. How a particular engine can be taken apart will be self-evident if you take the time to look at the construction. The nature of the problem also has a bearing on the extent of disassembly. The logical procedure is to limit the disassembly to only what you need to locate the problem and its cause.

REMOVE FLYWHEEL

With the cylinder head and blower housing removed, remove the flywheel. The flywheel is mounted on the tapered end of the crankshaft, held in position with a key and held on the crankshaft with a nut. Remove the nut. Use of a puller is recommended to loosen the flywheel. If a puller is not available, put the nut back on the shaft until it is flush with the end of the shaft. If the nut contacts the flywheel when it is flush, remove the nut. Carefully insert a heavy screwdriver between the block and the backside of the flywheel so you can exert outward pressure on the flywheel. Be sure the screwdriver is positioned so it does not cause damage when pressure is applied. Strike the end of the crankshaft a heavy blow with a rawhide or plastic hammer while prying outward on the flywheel. Figure 6-1 shows how to loosen the flywheel when a puller is not available.

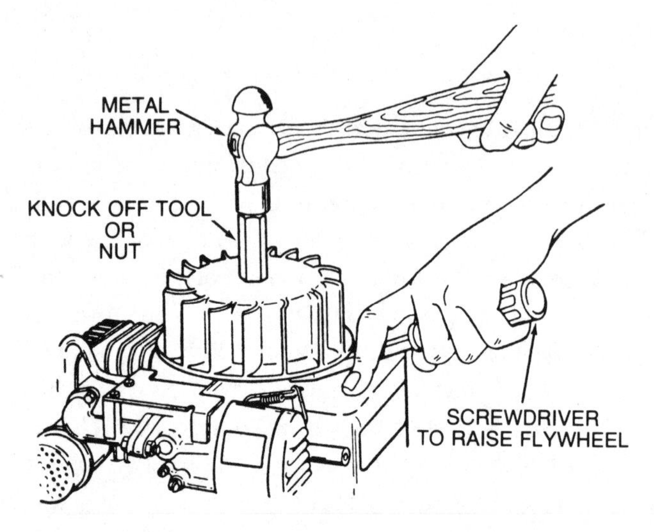

FIGURE 6-1
Removing flywheel
Courtesy Tecumseh Products Co.

If a heavy rawhide or plastic hammer is not available, place a piece of wood on the end of the crankshaft to protect the threads and strike the wood a heavy blow with a large hammer. This technique should loosen the flywheel. But take extreme care not to damage the threads on the crankshaft.

DISASSEMBLE CRANKCASE

How to take apart the major engine assembly is usually obvious. On some engines you can remove the cylinder barrel from the crankcase. Removing the cylinder barrel enables you to service the piston, rings, connecting rod, and cylinder barrel, as well as to check for crankshaft bearing wear. Figure 6-2 is a two-stroke cycle cylinder barrel removed from the crankcase.

Some engines utilize a two-piece crankcase that separates along the center, thus enabling you to service the crankshaft, connecting rod, and crankshaft bearings. Other engines may have a removable bearing support plate, which contains one of the crankshaft bearings (Figure 6-3). Still other engines may use two removable bearing support plates, one on each side of the crankcase. A few engines may have a removable oil base or sump, which should permit you to inspect and service the components located in the lower section of the crankcase.

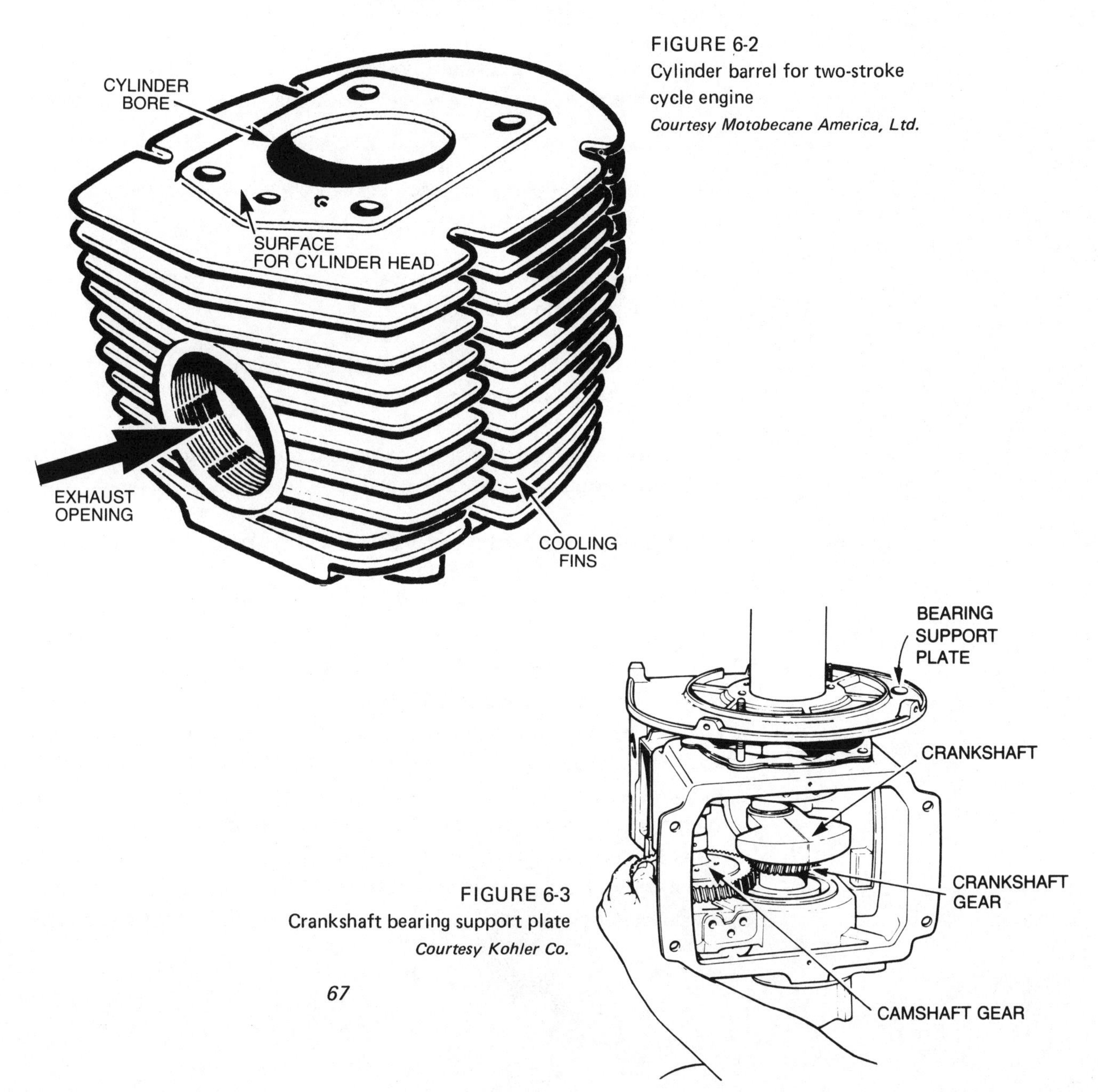

FIGURE 6-2
Cylinder barrel for two-stroke cycle engine
Courtesy Motobecane America, Ltd.

FIGURE 6-3
Crankshaft bearing support plate
Courtesy Kohler Co.

INSPECT INTERNAL ENGINE PARTS

With the engine disassembled to this extent, you should be able to determine where problems exist. At this point you need to make a decision as to the best method of taking care of the malfunction. Are you able to make the necessary repairs yourself? If so, can you get the needed parts? Do you have the necessary tools and equipment?

To give you an idea as to what may have to be done to restore the engine to normal performance, some of the more common repair jobs follow:

1. Should the cylinder bore (wall) be scored, worn, out-of-round, or grooved, it needs to be rebored and fitted with an oversize piston.

2. If the connecting rod bearing has excessive play (clearance up and down), a new connecting rod should be installed.

3. Should the crankshaft journals be worn, out-of-round, or scored, the shaft should be replaced or remachined and new bearings installed.

4. If the piston rings are broken or if they have lost their tension, they must be replaced.

5. If the piston is scored or if the ring lands are broken or burned away, the piston must be replaced.

6. Teeth on the crankshaft, camshaft, or governor gears may be chipped or worn and need replacing.

7. The spring on the automatic compression release may have become unhooked.

8. Oil seals may need replacing.

9. An exhaust valve seat insert may need to be replaced.

When the crankshaft and crankshaft bearing are damaged, if the cylinder wall is seriously damaged, or if you see some other severe damage, you might check on the availability and cost of a short block or mini-block. Some manufacturers define the short block as an engine assembly consisting of the valve mechanism, camshaft, piston and connecting rod, and crankshaft with bearings. All other items need to be transferred from the old engine. The mini-block, as sold by some manufacturers, includes the components that the short block has but excludes the crankshaft and bearing support plate.

SERVICE INTERNAL ENGINE COMPONENTS

The following information is relative to the complete servicing of the internal components of the engine, and it is general enough to apply to the majority of engines in common use. Exact specifications for specific engines will be found in the manufacturer's manuals. General specifications, clearances, tolerances, and the like, as presented, will ordinarily suffice.

A number of the service operations may not apply to your particular engine because it does not use the component discussed. Two-stroke cycle engines do not have mechanically operated valve systems including a camshaft. Some use a one-way reed valve or a rotating valve but most do not require any type of valve. Many engines do not have automatic compression releases, synchro-balances, or dynamic-balance counterweight systems on the crankshaft. Many low-horsepower engines do not use any of this equipment, and most do not use a mechanically driven governor. So the less equipment, the less complicated the repair operation.

Before attempting further disassembly, a word of caution: The relationships between certain parts are critical. The teeth on the crankshaft gear must be positioned in exactly the correct relationship to the teeth on the camshaft gear, because this positioning establishes valve timing. Examine the gears for the timing marks; when assembling the engine, you must line up the gear teeth exactly according to those timing marks. When rotating or oscillating counterbalances are used on the crankshaft, they must be reinstalled in the exact position from which they were removed. It is best not to disturb these units unless absolutely necessary. If you have any doubt about the markings for reassembly, make your own marks with a center punch.

In most cases, with the engine disassembled to this extent and the parts cleaned, you should be able to locate the trouble and determine what needs to be serviced or replaced.

The following is information about the servicing of the various internal engine components. The best way to approach the service operation is to visually inspect all parts for such abnormal conditions as wear (excessive free play), scores and scuff marks, breakage, cracks, or roughness. If there is a question about possible excessive wear, measure the component and check it against specifications, or compare it with a new part. A tester or an analyzer may be necessary to check ignition parts, or in some cases you can compare then with new units.

INSPECT CYLINDER BORE

Thoroughly clean the cylinder bore to remove all traces of carbon, dirt, oil, and sludge. Scrape off all old gasket material that may adhere to the cylinder block. Make a visual inspection for cracks, broken cooling fins, stripped bolts, or stripped bolt holes. Check the cylinder bore for grooves and scuff marks.

Check for a ridge or shoulder at the top of the cylinder wall. Scrape all traces of carbon from the top edge of the cylinder bore. The ridge, which results from wear, is located at the top of the cylinder bore down to the upper end of the piston ring travel area. It can be felt with your fingernail just below the top of the cylinder. If the ridge is too pronounced, the piston may be damaged when it is removed. The rings catch on the ridge and break the ring land. Use a ridge reamer to remove the ridge before removing the piston assembly. Figure 6-4 shows the use of a ridge reamer. Also, before removing the piston, check the connecting rod.

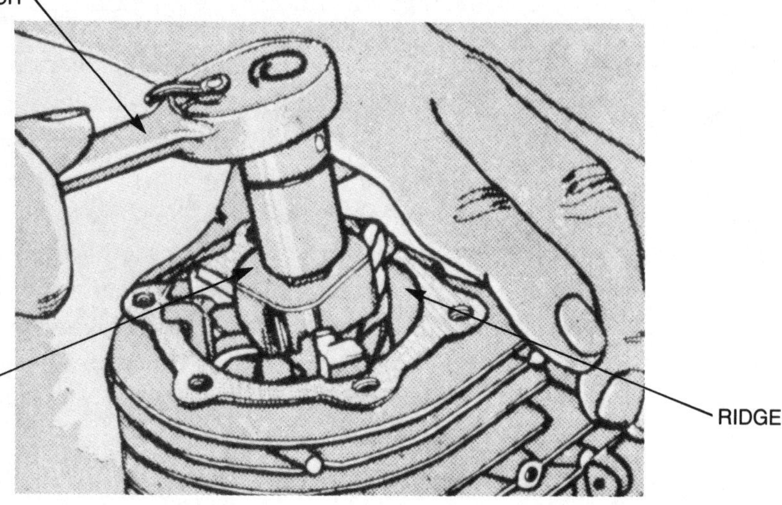

FIGURE 6-4
Removing cylinder ridge with a ridge reamer
Courtesy Clinton Engines Corporation

CHECKING CONNECTING ROD BEARING CLEARANCE

You have to remove the piston to completely inspect and measure the cylinder bore. Chances are you will also want to inspect the piston and rings, particularly if the cylinder bore is damaged in any way. The piston and connecting rod can be removed as an assembly from most engines.

Check the connecting rod for up-and-down movement on the crankshaft before loosening the bolts or nuts, if bolts are used to hold the cap to the connecting rod. To do so, grasp the connecting rod at the crankshaft and try to move it up and down on the crankshaft. There should be no noticeable free play up and down, but the rod will move freely from side to side. If there is no up-and-down movement, the connecting rod to crankshaft fit is all right. If there appears to be excessive free play (that is, wear) between the connecting rod bearing and crankshaft, then the connecting rod should be removed and replaced.

On some engines you have to take the crankshaft apart to replace the connecting rod bearing. The bearing is of the roller type.

An arbor press must be used to press the crankpin from the crankarm. Alignment is very important when reassembling a crankshaft of this type.

If the connecting rod cap and connecting rod bear no marks to distinguish how they fit together, use a center punch to mark both the cap and rod. The piston and rod assembly must be reinstalled in the cylinder bore in the same position from which they were removed, and the cap must be installed on the connecting rod in the same position.

Remove the connecting rod cap. The cap is held on the rod with bolts or screws. Cotter pins, self-locking nuts or screws, or locking tabs may be used. Remove the cotter pins or bend the locking tabs out of the way so as to be able to unscrew the bolts or nuts. Carefully push the piston and rod assembly out of the cylinder. If the ridge interferes, do not force the piston since it may crack or break the ring lands. Use a ridge reamer to remove the ridge. Figure 6-5 is a typical piston and connecting rod. The piston pin has been removed to separate the piston from the connecting rod.

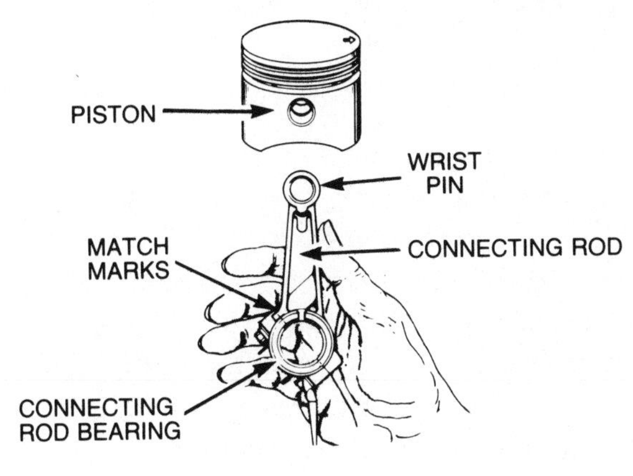

FIGURE 6-5
Piston and connecting rod disassembled
Courtesy Tecumseh Products Co.

Any evidence of scuffing, discoloration, grooves, or chipping on the bearing surface indicates that the connecting rod should be replaced. If the crankshaft and connecting rod appear to be serviceable, measure the clearance. The easiest method is to use a plastic gauge, which is a round piece of plastic about the size of the lead in a mechanical pencil. It comes in a paper wrapper which is used to measure the size of the flattened plastic in thousandths of an inch to indicate clearance. This gauge can be obtained from a garage or auto parts store. A piece about one inch long is all you need. Lay the piece of plastic on the crankshaft lengthwise to the shaft. Install the cap, and do not turn the crankshaft. Tighten the bolts or nuts that attach the cap to the connecting rod to approximately 120 inch-pounds. Carefully remove the bearing cap. Use the paper in which the plastic gauge came in to give you the operating clearance. Figure 6-6 shows how to check the bearing clearance with a plastic gauge.

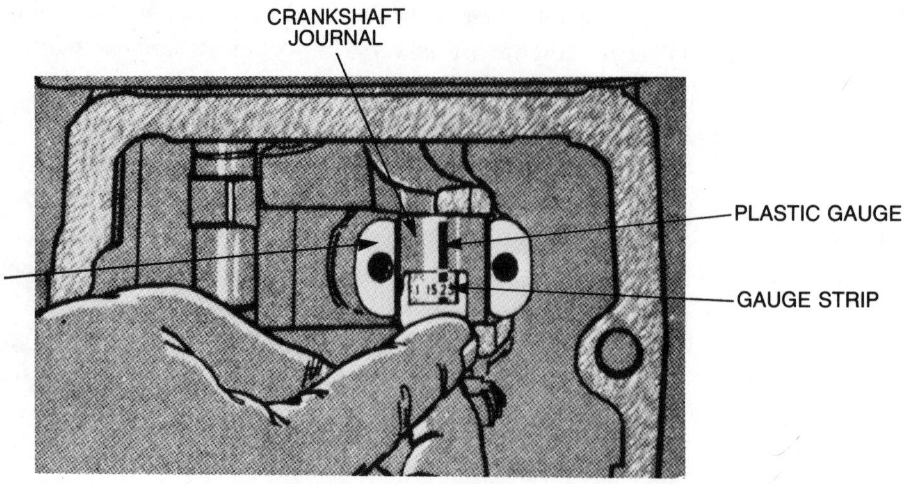

FIGURE 6-6
Checking connecting rod bearing clearance with plastic gauge
Courtesy Clinton Engines Corporation

To check crankcase "out-of-round," turn the crankshaft 90° and make another check with plastic gauge. The difference in the two readings indicates the out-of-round and shaft taper. The maximum allowable clearance (the wear limit) should not exceed 0.0035 inch. Out-of-round or taper should not exceed 0.0015 inch.

MEASURE CYLINDER BORE

Either an inside micrometer or a telescoping gauge with an outside micrometer is used to measure the cylinder bore. If the bore (the inside diameter) is oversized or tapered more than 0.005 inch, or if it is out-of-round more than 0.0025 inch, the cylinder should be remachined. Some engines come from the factory with an oversized piston; if so, the amount of oversize is stamped on top of the piston. To establish the original cylinder size when the specifications are not available, measure the top of the cylinder bore where there is no ring wear. Measure it with an inside micrometer or with a telescoping gauge and an outside micrometer. To check for wear, taper, and out-of-round, take a measurement just below the shoulder or ridge at the top of the cylinder at two points that are 90° apart. This reading indicates how much the cylinder is out-of-round and worn. Make the same type of measurement about 1½ inches from the bottom of the cylinder; this measurement indicates the out-of-round and wear at the bottom. A comparison of the top and bottom measurements shows the cylinder taper. Figure 6-7 shows the use of a telescoping gauge, which is used in conjunction with a micrometer, to check cylinder wear.

A dial indicator may also be used to show cylinder taper and out-of-round. Take several readings for an accurate indication.

The piston fit in the cylinder can be checked with a thickness gauge (feeler stock) and a tension scale. Place the piston upside down in the cylinder with the feeler gauge at right angles to the piston pin holes. With a tension scale, check the force required to pull the feeler

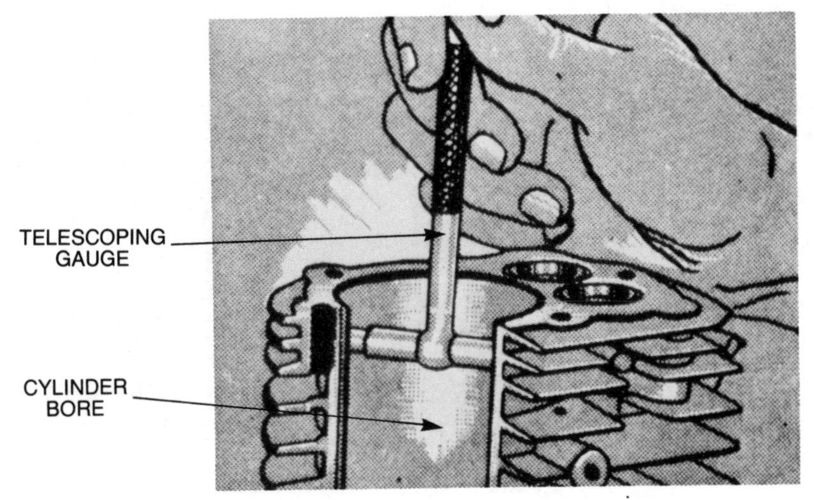

FIGURE 6-7
Checking cylinder bore with telescoping gauge
Courtesy Clinton Engines Corporation

gauge out. Some manufacturers recommend that it require from 6 to 9 pounds of pull to remove a 0.0015-inch feeler stock (gauge) that is ½ inch wide. Less than 6 to 9 pounds of pressure indicates excessive wear.

RECONDITION CYLINDER

If the cylinder bore is worn excessively, if it is badly scuffed, or if there are grooves in the wall, it should be remachined. If the cylinder is not worn extensively but has slight scuff marks or very shallow grooves, then fine emery paper can be used to remove the scuff marks and grooves. If new piston rings are going to be installed, use fine emery paper or fine wet or dry sand paper to lightly sand the surface. Sanding breaks the glaze on the bore so the new rings seat better. Do not break the glaze if the engine is equipped with an aluminum cylinder.

The cylinder bore is remachined with a hone. The process is sometimes referred to as "reboring the cylinder." The hone has adjustable grinding stones that are driven by a drill press or electric drill motor. Course abrasive stones may be used at the beginning to remove material, but fine stones are used to abtain a smooth finish when removing approximately the last 0.001 inch of material. The cylinder should be honed only enough to remove all grooves and scuff marks, as well as to remove the taper—in brief, only enough to provide a uniformly round and smooth surface. Figure 6-8 shows the hone being used in the cylinder bore. Whenever the cylinder is honed or the glaze removed, thoroughly clean the cylinder and block assembly with soap and water to remove all traces of abrasive.

When a cylinder is honed to remove any amount of material, an oversized piston must be installed. Pistons are available in the following oversizes: 0.010, 0.020, or 0.030 inch. The cylinder must then be honed to one of these oversizes. Do not remove any more material

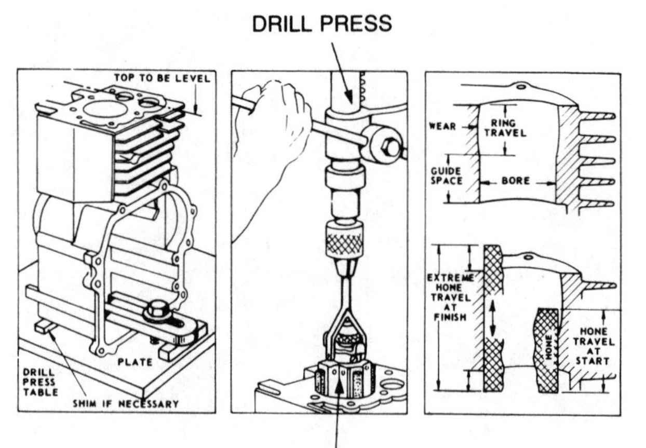

FIGURE 6-8
Honing cylinder bore
Courtesy Briggs & Stratton Corp.

than necessary to clean up the bore and to bring it to a standard required oversize. Since the necessary clearance for lubrication and expansion is provided for in the new piston, always obtain the oversize piston that is the same size as the cylinder bore.

CHECK PISTON

The function of the piston is to transfer the energy of the explosive force in the combustion chamber to the crankshaft. A good seal between the piston and cylinder wall prevents blow-by (leakage), which is the escape of gases from the combustion chamber into the crankcase and which causes loss of power.

Piston rings, which are flexible, are installed in grooves in the piston to prevent compression from escaping past the piston and to prevent oil in the crankcase from getting up into the combustion chamber.

If the cylinder is to be reconditioned, you need not check the piston because an oversized piston must be installed. New pistons ordinarily come with new piston pins. If the cylinder is not to be reconditioned and if the piston shows no signs of wear or scoring, check the piston for ring groove wear. If the ring groove wear, with a new ring, exceeds 0.007 inch, then the piston should be replaced. Figure 6.9 shows how to check the ring groove clearance.

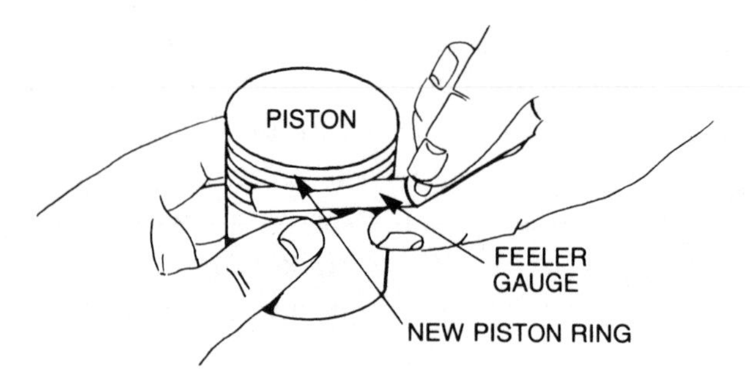

FIGURE 6-9
Checking ring groove clearance
Courtesy Briggs & Stratton Corp.

Remove the piston rings. Although there is a tool for this purpose, you should be able to clear the piston and remove the ring without breakage if you carefully spread the ends of the ring apart by using your thumbs. Always remove the top ring first. Some manufacturers recommend replacing rings whenever the piston is removed. This is a judgment call. If the rings still have tension, if they show no signs of excessive wear or scuffing, and if they do not have too much end gap clearance or groove clearance, they can be reused.

After removing the rings from the piston, clean all the carbon out of the ring grooves. Again, although there is a special tool for this purpose, you can use a piece of broken ring to remove the carbon. Whatever you use, be careful not to damage the ring lands (the material between the ring grooves). Make sure that the openings in the ring groove for the oil ring are open.

CHECK RINGS

Whether the rings are new or used, check the end clearance of the ring in the cylinder and the side clearance in the ring grooves. To check the end clearance, insert the ring in the cylinder bore. Push it down into the bore using the inverted piston so the ring is square with the cylinder wall. Push it down to about the center of ring travel. Measure the ring end gap with a feeler gauge (Figure 6-10).

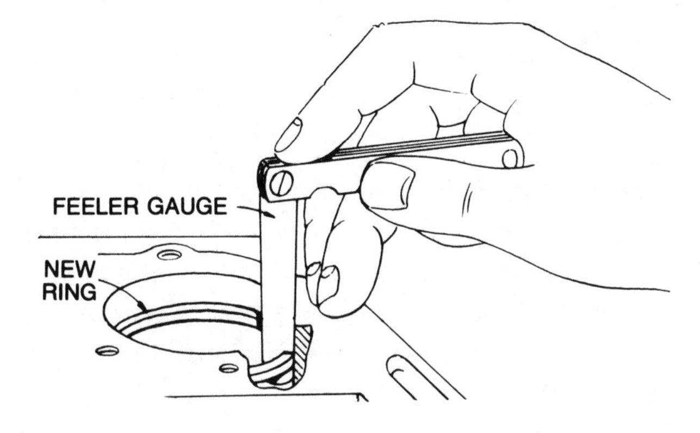

FIGURE 6-10
Checking ring end gap
Courtesy Briggs & Stratton Corp.

If the gap exceeds 0.030 inch, replace the ring. If the gap is less than 0.007 inch, carefully file the end of the ring to obtain a minimum of at least 0.007 inch. Check the side clearance with a feeler gauge after installing the ring in the proper groove. If the clearance exceeds 0.005 inch with an old ring, install a new ring and check it again. If the clearance is still in excess, replace the piston.

CHECK PISTON PIN

Check the piston pin for excessive wear in the piston pin bosses, as well as between the pin and the connecting rod if a bearing or bushing is used. Although normally very little wear takes place in this area, excessive wear may result in a slight rattling knock upon deceleration, but this does not cause damage.

To check for wear, grasp the piston in one hand and the connecting rod in the other. Try moving the connecting rod up and down to see if there is any free play between the piston and the piston pin and/or between the piston and the connecting rod. Side play will be present, but this has nothing to do with piston pin wear. If the piston pin is worn more than 0.001 inch or if the piston is worn more than 0.001 inch, then the problem is best corrected. The piston pin can be measured with a micrometer and checked against either a new pin or specifications. The hole in the piston can be measured with a telescoping gauge and micrometer. In most cases you have to replace the piston, piston pin, and connecting rod.

Some manufacturers make oversized piston pins available. If so, the piston pin bosses and connecting rod can be reamed or honed to fit an oversized pin.

Most piston pins are held in place by a retainer in the piston. Use a needle-nose pliers to remove the lock if the pin is to be removed. In some cases you may have to submerge the piston in hot water to expand the piston and thus to remove the pin.

If you checked the connecting rod before removing it and found it to be within wear limits, and if the crankshaft connecting rod journal is in serviceable condition, then the rings can be installed in the piston ring grooves. If the connecting rod is to be replaced, do so before installing the rings.

INSTALL RINGS

Piston rings are installed in sets and must be placed in the proper grooves. Despite the many variations of rings, you should have no trouble installing the correct ring in the proper groove, as long as you carefully observe the construction of each ring. Installation instructions are usually included with new ring sets. Make sure the piston and ring grooves are clean and free from all carbon and grit.

On four-stroke cycle engines, the oil control ring must be installed first. (Two-stroke types do not use an oil control ring.) The oil control ring may be made up of segments or of one piece. If one-piece, the ring has openings so the oil may return to the crankcase through the inside of the piston. If there is a spring expander included with the oil control ring, it must be installed in the groove before installing the ring. Do not spread the rings too wide when installing. If the oil ring has a bevel (chamfer) on the outside, then install the bevel toward the top. When the second ring has an inside chamfer, it goes up. When there is a notch on the outside diameter, the notch goes down; this notch acts as an oil scraper. If the top ring has a chamfer on the inside diameter, then the chamfer goes up.

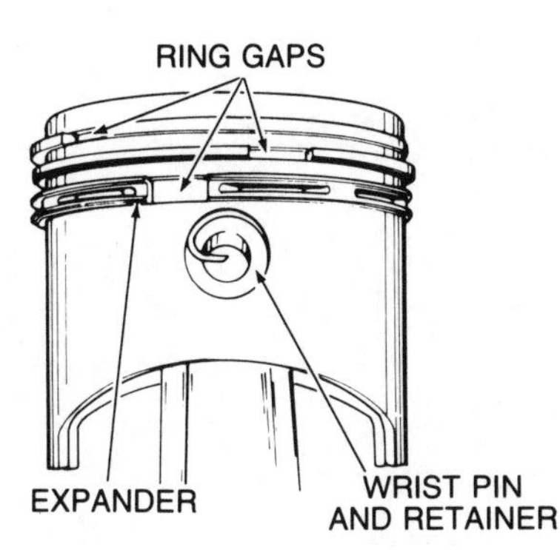

FIGURE 6-11
Rings installed on piston
Courtesy Tecumseh Products Co.

After the rings are installed on the piston, make sure they move freely in the grooves. Some pistons have a pin through the ring grooves so the rings cannot turn in the grooves. If so, turn all the rings so their ends are on either side of the pin. When a pin is not used, the rings should be turned so that the gaps are staggered (not lined up). Figure 6-11 shows the rings installed on the piston.

A ring compressor must be used to compress the rings into the grooves for installation purposes. If the engine is to be assembled at this time, carefully install the piston assembly into the cylinder bore after coating the piston with oil. If the piston is installed from the top, start the piston into the bore until it rests on the ring compressor edge. Tap the piston through the compressor into the cylinder. Be sure to hold the compressor tight on the cylinder so the rings cannot slip out of the grooves. When the cylinder barrel is installed onto the piston, carefully work the cylinder down over the rings, compressing each ring as you push the cylinder downward. Do not force the piston into the bore or the rings may be broken.

Install the connecting rod cap to the connecting rod if the cap was removed. The cap, rod, and piston must be installed in exactly the same relationship to the cylinder bore from which they were removed. If an oil dipper is used, this must also be installed in the correct position. Tighten the connecting rod bolts to the correct tension, which is approximately 125 inch-pounds. Bend up the locking tabs or install whatever locking device is used.

INSPECT CRANKSHAFT AND CRANKSHAFT BEARINGS

When should you check the wear factor of the crankshaft bearings? The answer depends to some extent on engine construction. If the bearing support is in a housing that must be removed to service the piston assembly, then the crankshaft bearing should be checked before disassembling the engine. To do so, try to move the crankshaft up and down by hand. A noticeable movement indicates bearing wear. Crankshaft end-play should be checked by trying to move the crankshaft in and out. Crankshaft end-play should not be excessive, not over 0.025 inch for most engines.

Two types of crankshafts are in common use: (1) the shaft that is cast as one piece and (2) the fabricated three-piece shaft.

Of importance is the manner in which either is serviced. With a *one-piece crankshaft*, the connecting rod has a removable cap, and the connecting rod and piston assembly can be removed without disturbing the crankshaft. Figure 6-12 shows a typical crankshaft and camshaft installation. If the bearing surface of the connecting rod is scored, worn, or chipped, then the complete connecting rod assembly should be replaced.

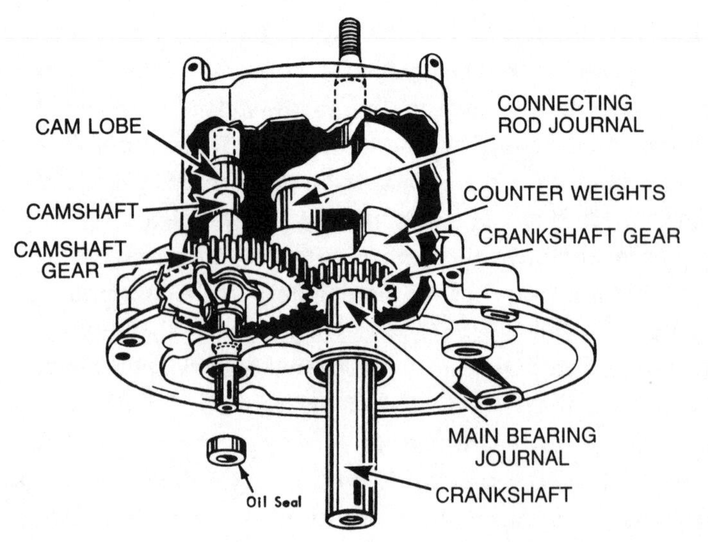

FIGURE 6-12
Typical crankshaft and camshaft installation
Courtesy Clinton Engines Corporation

If the crankshaft journal (bearing surface) is worn, scored, tapered, or out-of-round, it should be replaced or the journal remachined. Use an outside micrometer to check the wear condition of the crankshaft.

A *three-piece fabricated shaft* must be disassembled to service or remove the connecting rod. The connecting rod journal can be replaced if necessary, as well as either end of the crankshaft. Figure 6-13 shows a typical three-piece fabricated crankshaft. The end of the shaft that operates the magneto is sometimes called the "magneto end," and the end that drives the equipment may be labeled the "power take-off end."

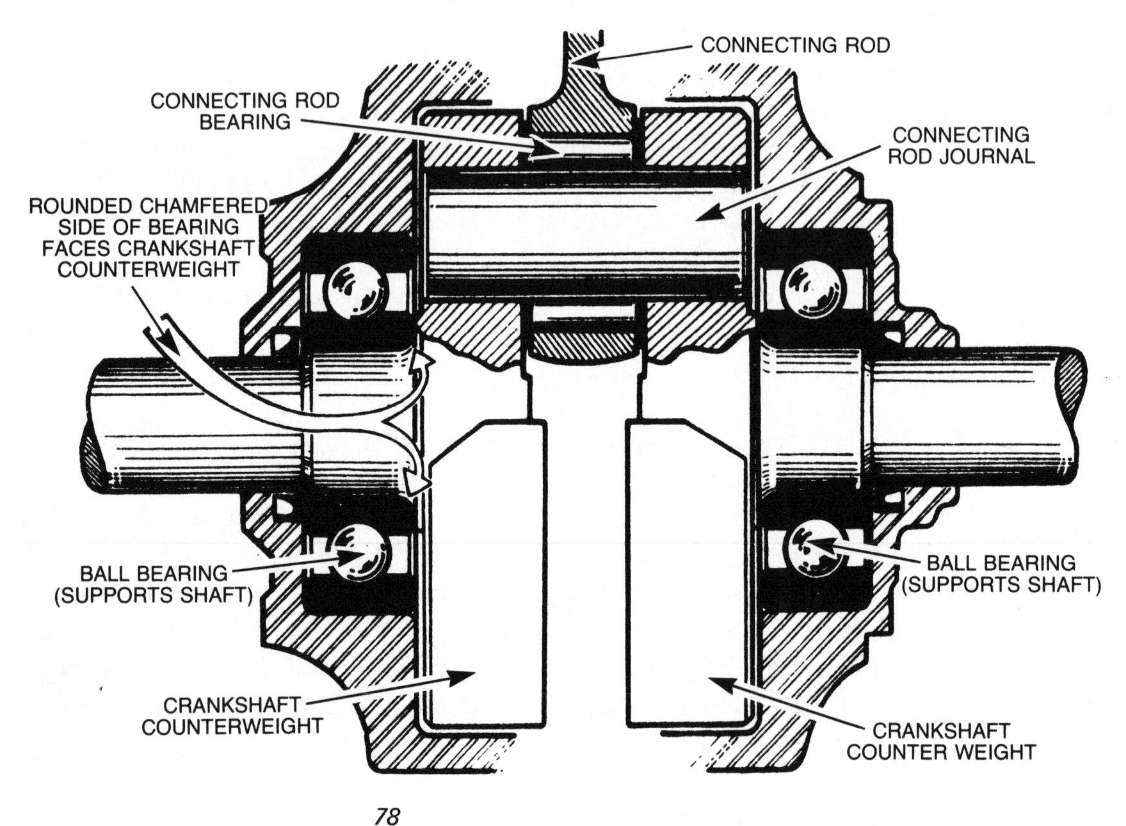

FIGURE 6-13
Three-piece crankshaft
Courtesy Motobecane America Ltd.

Whenever crankshaft and/or crankshaft bearing problems occur, you are generally better off either replacing the entire lower section of the engine or taking the unit to a small engine repair shop. To disassemble and reassemble the crankshaft requires an arbor press and special jigs. You also need a bearing puller to remove the ball or tapered bearings from the shaft. Whenever the crankshaft is removed, new oil seals should be installed to assure no leakage, and an oil seal driver is generally necessary to install oil seals properly.

If bushings are used to support the crankshaft and if these bushings are worn, new ones must be pressed into the housing and reamed to the correct fit. Washers of varying thicknesses are commonly used next to the bushings or bearings to establish crankshaft end-play. When removing the shaft, pay attention to the spacer washers so you get the correct end play when reassembling. Figure 6-14 illustrates the removal of the crankshaft support bushing.

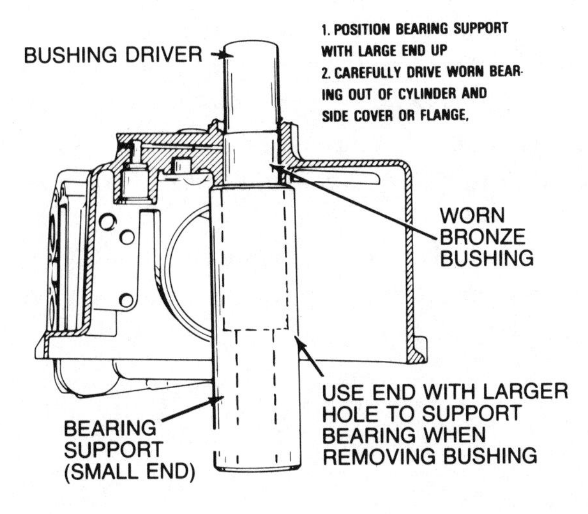

FIGURE 6-14
Removing crankshaft support bushing
Courtesy Tecumseh Products Co.

On most engines, the crankshaft is supported on ball bearings. Some may use roller bearings, while others use bushings. When the shaft is supported on ball or roller bearings, if properly lubricated, very little wear takes place on either the shaft or bearings; so the crankshaft and crankshaft bearings normally require very little service, if any. Figure 6-15 shows a crankshaft mounted on ball bearings. If inspection indicates no excessive wear (free play) between the bearings and crankshaft and if the shaft turns over smoothly with no roughness, check to make sure the threads on the shaft are not damaged and that the keyway is not spread or worn. Check to make sure the teeth on the crankshaft gear does not show signs of wear. Remove any burrs, nicks, or scratches on the shaft or gear with crocus cloth. It is best not to remove the crankshaft unless service is required.

In a four-stroke cycle engine, a gear mounted on the crankshaft is in mesh with a camshaft gear. The camshaft gear is twice the size of the crankshaft gear so the camshaft turns at half the crankshaft speed. Since the function of the camshaft is to open the valves, the

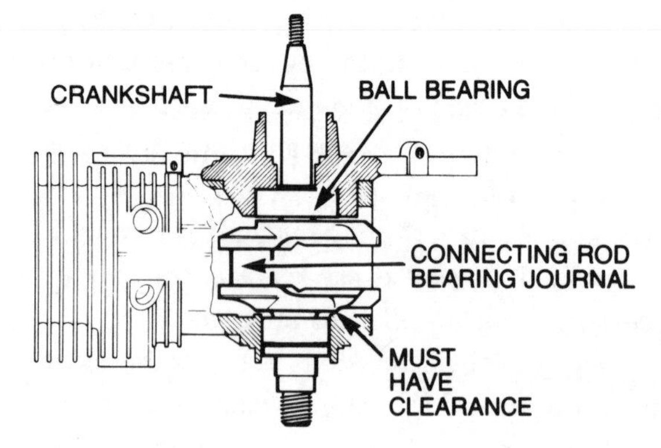

FIGURE 6-15
Crankshaft supported on ball bearings
Courtesy Tecumseh Products Co.

camshaft and the crankshaft must have a fixed relationship to one another. This relationship controls the valve timing.

The camshaft and crankshaft gears are marked in some manner to enable you to check and to reestablish timing should the gears or shaft be removed. Identify the marks and their relationship before removing the shaft or gear. There may be a number of marking devices: an O or I on both gears or just on one gear; lining up the keyway with the unmarked gear; a bevel on a gear tooth. Figure 6-16 shows typical timing marks found on the crankshaft and camshaft gears. If you are unable to locate any such marks, make your own using a center punch.

On some engines the crankshaft can be removed without disturbing the camshaft. Other engines require that the camshaft be removed in order to remove the crankshaft.

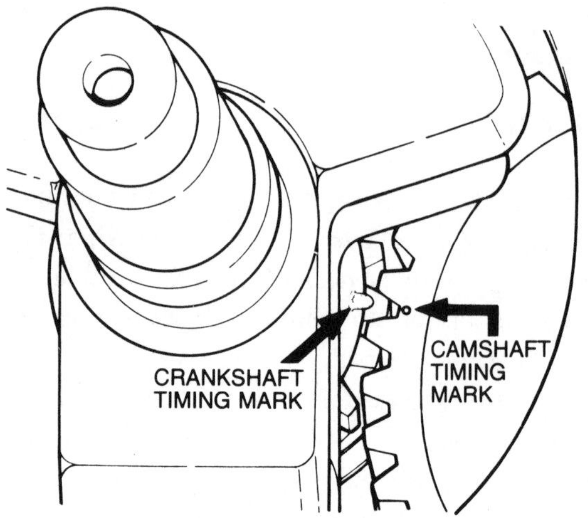

FIGURE 6-16
Timing marks on crankshaft and camshaft
Courtesy Kohler Co.

INSPECT CRANKSHAFT BALANCE SYSTEM

Different balance systems are found on different engines. Some of the larger engines use a dynamic balance system to reduce engine vibration. Of major importance is that, whenever disassembly is anticipated, the counterweights must be exactly synchronized with the crankshaft. Always locate the timing marks or establish your own using a center punch to mark the counterweights and gears, so that the exact relationship can be established for reassembly.

1. In one type of balance system, synchronized balance gears are mounted on stub shafts and turn on needle bearings (Figure 6-17). The gears are in mesh with one another and driven by a gear on the crankshaft. The stub mounting shafts are pressed into bosses in the crankcase. Snap rings hold the gears in place, and spacer washers control end-play. Timing marks are found on the gears. Become familiar with the line-up of the marks so you are able to reinstall the gears in the proper position. A special tool simplifies the timing process for this type of gear arrangement. If you must remove the gears and/or shaft, make a sketch of how the marks relate to one another; this diagram helps you when assembling the unit.

If the gear teeth are badly worn or chipped, or if the shaft that the gears revolve on is worn, then the balance unit must be removed and the gears, pins, and/or bearings replaced.

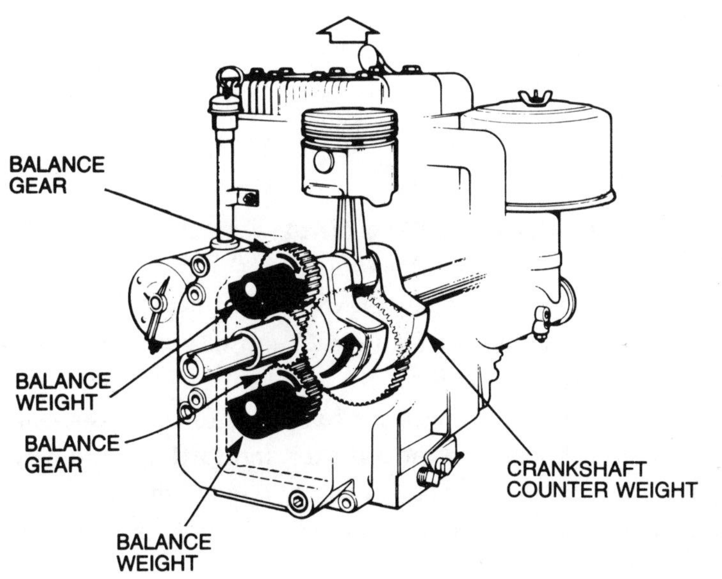

FIGURE 6-17
Synchronized balance gears
Courtesy Briggs & Stratton Corp.

2. Another type of dynamic balance system, the synchro-balance system, uses a rotating counterbalance. The counterweights are geared to rotate in a direction opposite to crankshaft rotation.

3. A third system uses counterweights that oscillate opposite to the direction of the crankshaft.

Little or no service should be necessary for any of these balance systems unless the gears become badly worn, the gear teeth chipped, or the bearings worn. If you have to remove the crankshaft, just be sure to establish the correct position of the gears that drive the weights.

INSPECT CAMSHAFT AND CAMSHAFT GEAR

As the camshaft turns at only one-half the crankshaft speed, there should be very little wear. Nonetheless, when the crankcase is disassembled, check the camshaft gear for worn, nicked, or chipped teeth, as well as for any freeplay in the bearings.

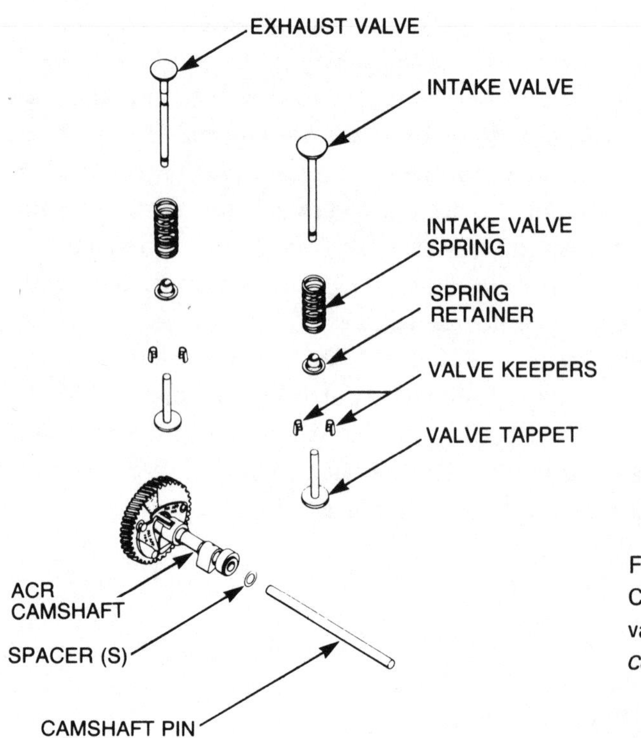

FIGURE 6-18
Camshaft, camshaft gear, and valve mechanism
Courtesy Kohler Co.

Inspect the camshaft lobes for signs of wear. If you find any, replace the camshaft. If the valve tappets (push rod), located between the camshaft and valve stem, shows signs of wear, remove the camshaft and replace the tappet. Figure 6-18 shows a camshaft, camshaft gear, and the valve mechanism.

Check for free play and end-play by trying to move the camshaft up and down for bearing wear and back and forth for end-play. If you get any indication of excessive play, remove the shaft. The removal will be obvious. The camshaft gear is normally pressed onto the shaft and retained by a cap screw or bolts.

You may find an engine that has a set of weights on the inner surface of the camshaft gear (Figure 6-19). This is an automatic spark

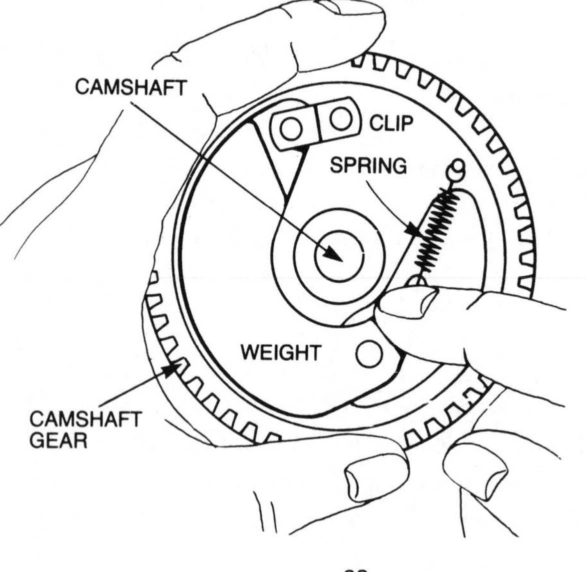

FIGURE 6-19
Camshaft with centrifugal spark advance
Courtesy Briggs & Stratton Corp.

advance. The camshaft hub is attached to the weights. As speed increases the weights move outward, turning the camshaft and advancing the ignition timing. Inspect the weights for freedom of movement and the springs for enough tension to lift the weights. If the weight mechanism is worn, replace the assembly including the camshaft and gear assembly. If the springs do not have that tension, replace them.

INSPECT AUTOMATIC COMPRESSION RELEASE

The automatic compression release mechanism is incorporated with the camshaft and camshaft gear. If the release uses a plunger (a release pin) and shows signs of wear, the entire camshaft must be replaced (Figure 6-20).

Different types of release mechanisms may be used on some installations, particularly on the higher-horsepower engines. One type of compression release mechanism uses flyweights on the camshaft gear. Although very little wear takes place, the spring that holds the weights in a released position may become unhooked. In that case, the spring has to be rehooked or replaced. If the flyweights become permanently stuck or excessively worn, a new camshaft should be installed.

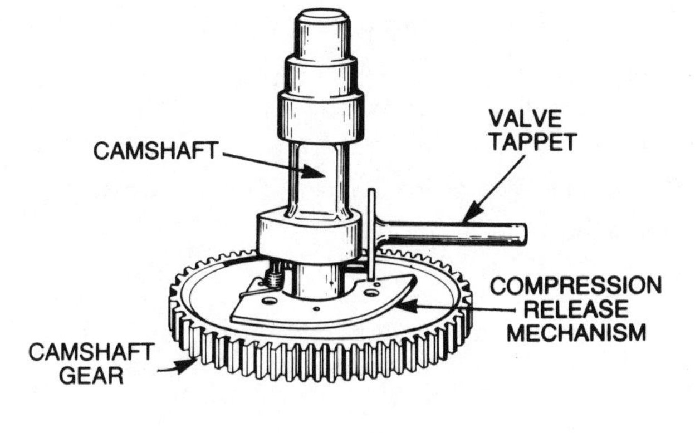

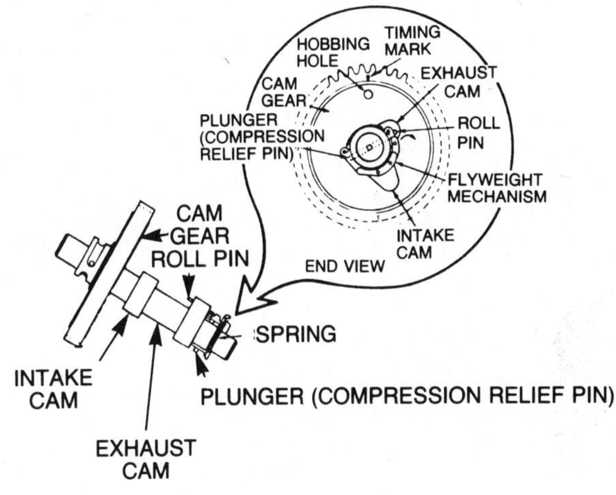

FIGURE 6-20
Compression release mechanism
Courtesy Tecumseh Products Co.

INSPECT MECHANICAL GOVERNOR

Most engines are equipped with either a mechanically driven governor or an air vane governor. The mechanical governor drive gear is located inside the crankcase and driven by the camshaft (Figure 6-21). The gear should be checked whenever the engine is disassembled enough that you can inspect the governor drive gears and the weights. The flyweights, which are turned by the governor drive gear, move outward due to centrifugal force as the engine speed increases. Spring tension pulls the flyweights inward as speed is reduced.

A cross-shaft actuated by the movement of the flyweights extends to the outside of the crankcase. An arm is attached to the

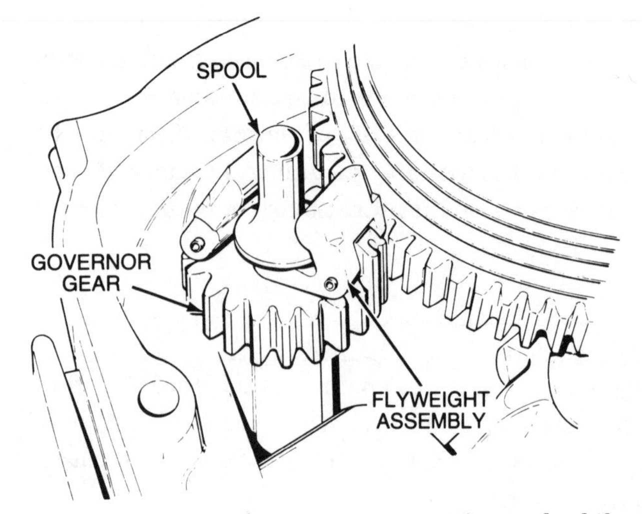

FIGURE 6-21
Mechanical governor drive gears
Courtesy Tecumseh Products Co.

outer end of the shaft, and a linkage connects the arm to the throttle valve in the carburetor. The governor is adjusted at the factory and should require no further adjustment unless the arm comes loose or the linkage is worn.

The purpose of the governor is to maintain engine speed at a relative constant rate. If you have to remove the arm, install it in the exact position from which it was removed; otherwise operation will be affected.

Inspect the gear for nicks, chips, or worn teeth. The flyweights must move freely on the pivot pins without excessive free play. The governor spring must not be stretched or distorted.

Some engines use a governor contained in a housing. The housing can be removed from the engine without disassembling the rest of the engine. If the governor gear or flyweights are worn, the entire unit should be replaced.

INSPECT OIL PUMP DRIVE GEAR

Some engines are equipped with a gear-driven oil pump. Some use a plunger-type pump, while others may use an oil slinger to direct internal engine lubrication. A great many engines, particularly low-horsepower ones, do not use any type of oil pump. Whatever the type of pump, it is driven by the crankshaft or camshaft.

The servicing of the various oil pumps, including the complete lubricating system, is covered in Chapter 7, "Lubricating Systems." Suffice it to say for now that, if you suspect trouble in the lubrication system, check and service it while the engine is dismantled. Replace the gear that drives the pump, as well as the pump-driven gear if the teeth are chipped, nicked, or worn. If the lubrication system has been functioning properly, it should not be disturbed. When a pump is not functioning properly, the entire unit should be replaced.

INSTALL OIL SEALS

To prevent oil from escaping on four-stroke cycle engines, oil seals are used where a shaft extends to the outside of the engine. In two-stroke cycle types, the seals prevent the air-fuel mixture from leaking past the shafts. Of major importance are the seals at either end of the crankshaft. Since the shaft revolves, it must have some working clearance, and so good seals are essential. An oil seal is used at the output shaft of the power take-off or gear reduction unit, if the engine is equipped with such a component. The governor shaft, which extends through the crankcase, may also require an oil seal.

Any time an engine is dismantled, replace the oil seals so as to be assured of leak proof joints. The seals are not expensive, but they must be handled and installed with care. A special oil seal puller and driver make the job easier. Figure 6-22 shows how to use a driver to install a new seal.

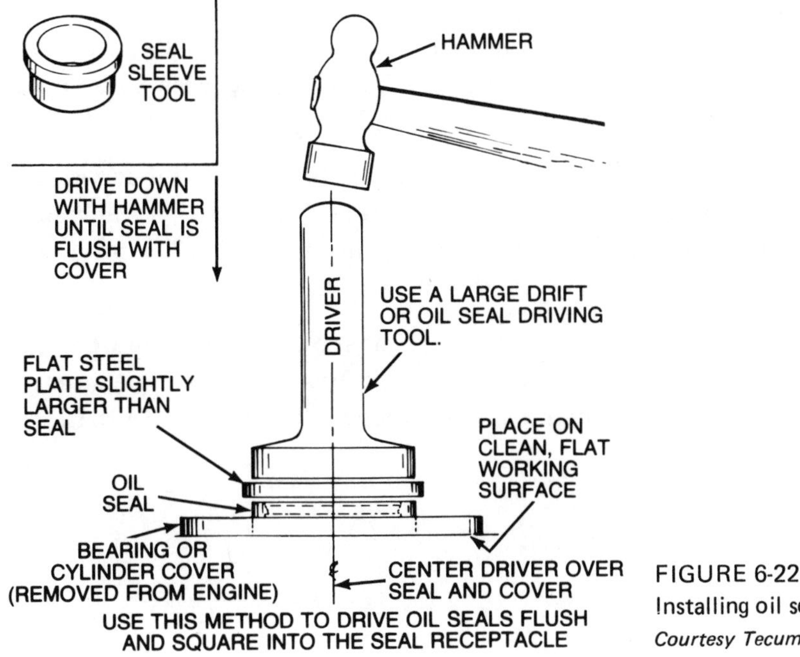

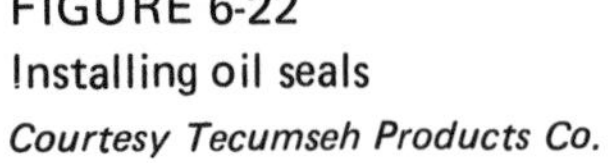

FIGURE 6-22
Installing oil seals
Courtesy Tecumseh Products Co.

If a seal puller and driver are not available, you can, with care, drive the old seals out of the housing without marring the surface. When installing new seals, it may be possible to use the old seal to drive the new one into place. The lip on the inside of the seal must face the inside of the engine. Put gasket sealer on the outside of the new seal to reduce the chances of leakage between the outside of the seal and the case. Place the old seal on top of the new seal. Lay a small piece of wood on top of the old seal, and tap on the wood to drive the new seal into place. Make sure the new seal goes in straight and seats tightly.

ASSEMBLE ENGINE

When assembling the engine, consider the following tips:

1. First make sure that everything is ready to assemble, that all bolts are properly tightened, and that, if locks are used, they are in place.

2. Apply a coating of oil to the moving parts to assure lubrication until the engine is started and running.

3. Make certain all matching surfaces are clean and free from old gasket material, dirt, and grit.

4. Determine whether a unit is serviceable by the amount of free play (clearance). All moving parts should move freely without binding, drag, wobble, or misalignment, a slight amount of clearance must exist. There can be some side-to-side movement (end play), but this must be limited.

5. In addition to replacing oil seals, use new gaskets throughout the engine. Gaskets can usually be purchased in sets. Placing a coating of heavy grease on the surface to which you wish to apply a gasket helps to hold the gasket in place during assembly.

6. During assembly, make sure all the parts line up without forcing. Where there are alignment marks, such as those used on the crankshaft and camshaft gears and on the connecting rod and bearing cap, make sure they are correctly positioned.

7. When bolting units together, tighten all the bolts uniformly, a little at a time, to prevent warping and breakage. A torque wrench should be used to tighten the bolts evenly and to the correct tension.

7 ENGINE LUBRICATION

ENGINE OILS
TYPES OF OIL DISTRIBUTION SYSTEMS
OIL PUMPS
GEAR PUMPS
BARREL-AND-PLUNGER OIL PUMPS
TWO-STROKE CYCLE ENGINE LUBRICATION
CHAIN SAW LUBRICATION

Maintaining the proper lubrication of engine parts is of vital importance. Without lubrication, the engine soon fails to function.

Engine oil does several things. It reduces friction, which causes heat and wear, by maintaining an oil film between the moving parts. Oil serves to dissipate heat within the engine. It acts as a cushion to reduce the shock of the load between the bearings and shafts. Oil forms a seal between the piston and cylinder wall to help prevent the loss of compression from blow-by. The oil cleans the metal parts that it comes in contact with and holds dirt, carbon, and metallic particles in suspension. In general proper engine lubrication reduces engine wear and makes for a smoother-running engine.

Load-carrying moving parts must have lubrication between the friction surfaces. The cylinder wall must have an oil film at all times, and the connecting rod bearing and piston pin bearing must also be lubricated. The crankshaft bearings likewise must receive lubrication at all times, as well as the camshaft and other moving parts.

The volume of oil must be such that all necessary parts are lubricated, but with too much lubrication, some of it may leak past the oil seals. Too much oil can also flood the cylinder wall and get past the piston rings into the combustion chamber where it is burned. This flooding can result in a fouled spark plug and cause the engine to emit blue exhaust smoke. The engine oil should always be kept at the recommended level for best performance.

Engine oil can be lost in different ways: It can leak past oil seals and gaskets, burn up in the combustion chamber, or escape as a vapor through the breather. Worn oil seals permit oil to escape around the shaft to the outside. The same happens if crankcase gaskets are cracked and torn, or if the matching surfaces that the gaskets are to seal are not properly tightened.

Worn piston rings or cylinder walls may permit oil to escape past the piston into the combustion chamber. The heat in the combustion chamber causes the oil to burn. A plugged breather system can result in pressure build-up within the crankcase. This pressure forces oil past the piston rings or out through the breather. Excessive blow-by can also force oil vapor out through the breather opening.

ENGINE OILS

Engine oils are classified according to the type of service and viscosity. The common classifications that are acceptable for most small gasoline engines include SC, SD, SE, or MS. The markings are found on the oil container. These oils contain a detergent that helps to keep the internal parts of the engine clean.

A special two-cycle oil is used in a two-stroke cycle engine. This oil does not contain a detergent. In addition many two-cycle oils contain no ash-forming metallic additives. These additives may cause pre-ignition problems, which result in scoring and scuffing of the piston and cylinder walls. Most two-cycle oils contain a rust inhibitor to help protect engine parts and a diluent to improve the mixability of the oil with gasoline, especially at low temperatures.

The present oil-to-gasoline ratio has been improved to the point that most manufacturers recommend a 50-to-one fuel-to-oil ratio. The advantage of this development is that having as little as possible oil per tankful is more economical, and you do not lose power, give off as much smoke, or foul spark plugs.

High-quality two-cycle oil offers improved lubrication, reduced combustion chamber deposits, less piston ring sticking, minimized piston skirt deposits, and longer spark plug life. It contains a rust inhibitor, has good mixibility with gasoline, and reduces the chances of pre-ignition.

When filling the fuel tank on a two-stroke cycle engine, put in about half the gasoline, add all the oil, and then put in the remainder of the gasoline. If the fuel and oil are being put into a separate can, which is the best way, shake the can well to mix the fuel and oil before putting the fuel mixture in the tank.

The viscosity of an oil is important in that it indicates its thickness: The higher the viscosity, the greater its thickness. When oil is cold, it is thicker and pours more slowly than when hot. When the weather is cold, the engine turns over much harder when an oil with a high viscosity number is used. On the other hand, in hot weather an oil with a low viscosity number may become so thin that it fails to lubricate properly.

You can see from all this that it is important to follow the manufacturer's recommendation as to the grade, amount, and type of oil to use. It is also essential that the oil be changed according to recommendations from the manufacturer.

While oil does not wear out, other things happen to it. It picks up dirt from the air, as well as sludge and carbon from within the engine. Water may accumulate in the oil due to condensation. Gasoline may also get past the piston rings into the crankcase, particularly if the engine has been flooded while starting.

When the manufacturer's recommendations are not available, the following rules may serve as a guide. Most manufacturers recommend changing the oil every 25 hours of operation; when operating under extremely dusty conditions, change it more often. Generally SAE 30 viscosity oil is satisfactory when temperatures are above 32°. When they are below 32°, an SAE 10W oil should be satisfactory. Most manufacturers approve the use of a multiviscosity oil such as 10W-30. This grade does the trick under most operating conditions.

TYPES OF OIL DISTRIBUTION SYSTEMS

Oil must be distributed within the engine so that all the parts needing lubrication receive the required amount. The volume of oil reaching certain areas must be controlled. The oil seals can retain only a certain amount of oil. In too great a quantity, the oil leaks past the seals. If too much oil gets on the cylinder wall, the rings are not able to return the oil to the crankcase, and it gets past the piston rings into the combustion chamber.

Many small low-horsepower engines rely simply on a splash-type lubrication system. A dipper, generally a part of or attached to the connecting rod cap, directs the oil onto the moving parts by dipping into the oil reservoir. Figure 7-1 illustrates an oil dipper that is part of the connecting rod bearing cap. It is important that, if the dipper is removed from the connecting rod cap, it is reinstalled in the correct position, so as to pick up the oil in the proper manner.

Some engines use an oil slinger—in the form of a wheel with teeth—to better distribute the oil. It is mounted on a bracket on the end of the camshaft and driven by the camshaft gear. The teeth on the wheel pick up oil and distribute (sling) it onto the moving engine parts. If any of the oil distribution parts, such as the bracket, gear, or slinger, shows signs of wear, replace the entire unit.

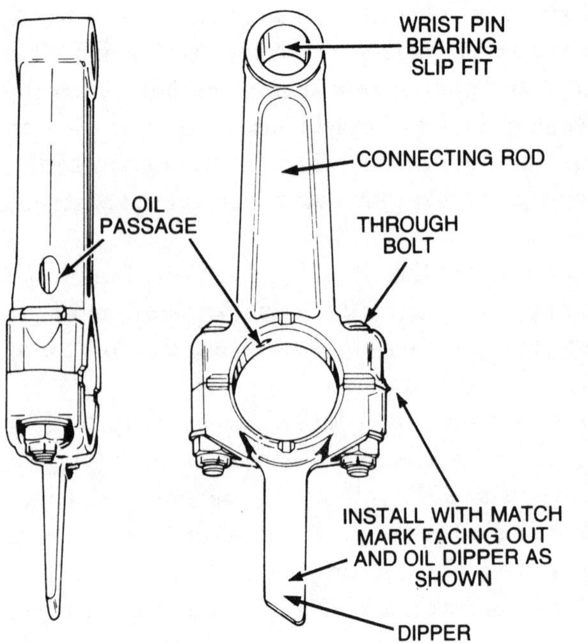

FIGURE 7-1
Oil dipper, an integral part of connecting rod cap
Courtesy Tecumseh Products Co.

Another type of slinger, attached to the crankshaft, scoops up the oil as the shaft turns. Figure 7-2 shows this type of oil slinger. As you can see, it is a very simple device that should not wear out.

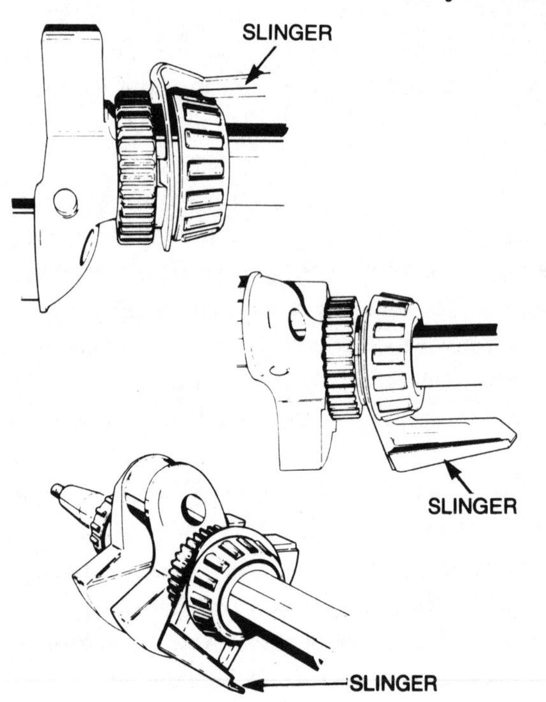

FIGURE 7-2
Oil slinger on crankshaft
Courtesy Tecumseh Products Co.

OIL PUMPS

Oil pumps are used in some engines, particularly those that have a vertical crankshaft. The pump, which may be of the gear-type or of the barrel-and-plunger type, provides a positive supply of oil. Both types are driven by the camshaft.

GEAR PUMPS

The gear-type pump draws oil from the oil sump (reservoir) in the engine base and forces it to the camshaft through a drilled camshaft passage to the top camshaft bearing. From the camshaft bearing oil flows to the top main bearing, then through a drilled passage in the crankshaft to the connecting rod bearing. The oil that is forced out from around the connecting rod bearing lubricates the cylinder wall. Normal oil splash lubricates the rest of the internal engine parts. A pressure relief valve is used in conjunction with the pump to relieve excessive pressure when the oil is thick and cold, as well as at high engine speeds.

The pump is made up of a drive gear and a displacement member (Figure 7-3). As the drive gear turns, it creates pressure because the oil is squeezed between the drive gear and displacement member.

FIGURE 7-3
Gear-type oil pump
Courtesy Clinton Engines Corporation

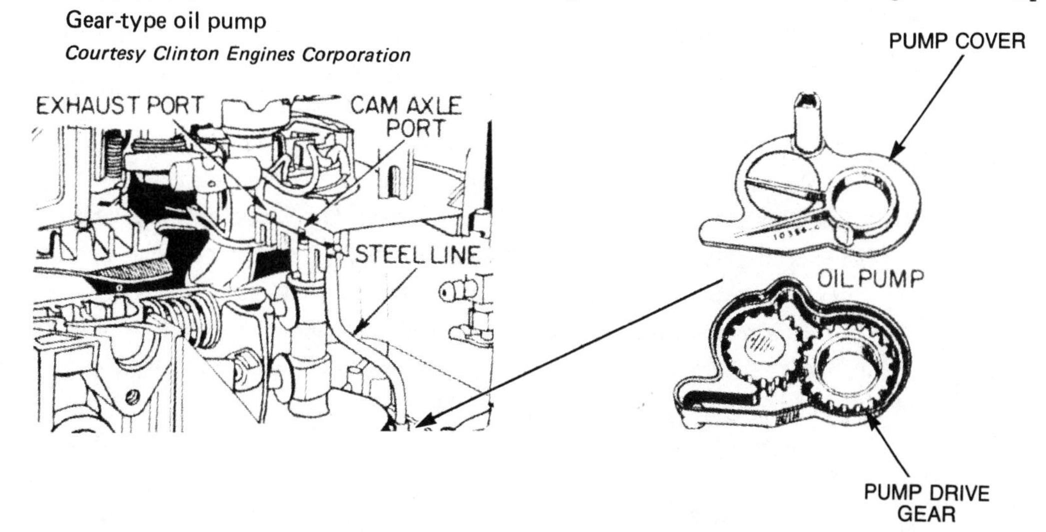

The pump normally operates satisfactorily over a long period of time and should not have to be disassembled. If necessary, however, the pump can be disassembled by removing the screws that attach the cover. If the pump is disassembled, inspect the pump drive gear and displacement member for wear, scoring, broken teeth, or other damage. Inspect the case and cover for scoring, wear, or cracks. The entire pump assembly should be replaced if the parts show signs of wear or damage. The cover gasket provides clearance for the drive gear. Check the clearance between the cover and the oil pump gear with a feeler gauge. The clearance should be 0.006 to 0.007 inch. Different gasket shims are available to obtain the correct clearance. Fill the pump cavity with oil before assembly.

BARREL-AND-PLUNGER OIL PUMPS

The barrel-and-plunger oil pump is driven by a lobe on the camshaft (Figure 7-4). On the intake stroke the pump draws oil through the hollow camshaft from the oil sump (reservoir). The passageway from

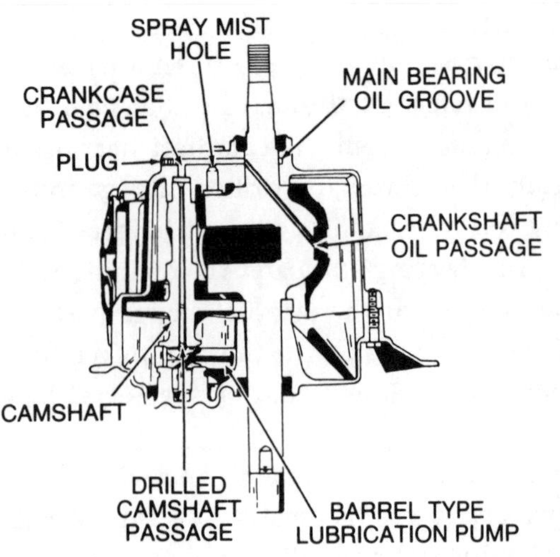

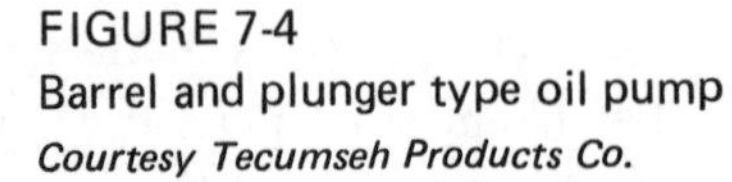

FIGURE 7-4
Barrel and plunger type oil pump
Courtesy Tecumseh Products Co.

the sump through the camshaft is aligned with the opening in the pump. As the camshaft continues to rotate, the plunger forces oil out of the top of the camshaft. From the top of the camshaft oil is forced through a passage in the crankcase to an oil groove in the top main bearing. From there the oil is forced through a passage in the crankshaft to the connecting rod journal. Oil escaping from around the connecting rod bearing lubricates the cylinder wall. Normal splash lubricates the other internal engine parts.

A pressure relief valve, located in a crankcase passage, is used to relieve excessive pump pressure. This is a simple spring-loaded valve that lifts off its seat when pressure becomes too high. On some models the pressure relief port is designed in the form of a spray mist orifice.

The lower main bearing receives lubrication through a groove in the bearing.

The camshaft must be removed to remove the pump. Separate the barrel and plunger, and then inspect them for rough spots or wear. If either is rough or worn, replace the entire pump.

When assembling, fill the barrel with oil and install the plunger. Operate the pump manually to make sure the plunger slides freely.

Install the pump on the camshaft. The chamfer side must be placed toward the camshaft gear. Install the mounting flange. Be sure the plunger ball seats in the recess in the flange before tightening the flange to the cylinder.

TWO-STROKE CYCLE ENGINE LUBRICATION

The lubricating oil is mixed with the gasoline in a two-stroke cycle engine. The air-fuel mixture, which contains a specified amount of oil, is drawn into the crankcase from the carburetor and up a passageway to the intake port. For this reason it is not possible to maintain a reservoir of oil in the crankcase.

The air-fuel-and-oil mixture is drawn into the crankcase. Since the air-fuel part of the mixture is more volatile than the oil, it evaporates and passes on to the combustion chamber as a combustible mixture where it is burned. The remaining oil is enough to maintain an oil film on the moving parts.

The ratio of oil to gasoline may vary among makes and models of engines, as well as with the particular oils used. A decal near the fuel tank filler opening normally describes the mixture to use. Always follow the manufacturer's recommendations.

CHAIN SAW LUBRICATION

The power plant used for a chain saw is a two-stroke cycle engine, so the engine lubricating oil is mixed with the gasoline. The cutting chain, however, has a separate, pump-operated lubricating system. The service life of the chain and the guide bar depends to a great extent on their proper lubrication, as well as on keeping the chain sharp and snug in the guide bar.

Every time the saw is refueled, the oil tank should be topped off with oil. A special chain lubricating oil should be used to lubricate the chain. When an approved chain oil is not available, a good quality single-grade engine oil can be used. Use SAE 30 lubricating oil if the lowest temperature is above 40°F. If the temperature is lower than 40°F, use SAE 10.

The oil pump, the plunger-type, is usually driven by the chain sprocket, so the pump operates only when the clutch is engaged. On most installations an adjustment screw is located on the outside of the pump body to permit the operator to increase or decrease the amount of oil delivered to the chain. Some pumps are equipped with a hand-operated pump, which is incorporated into the engine-driven pump. Pushing the exterior knob on the pump produces additional oil for the chain before starting the engine.

Figure 7-5 shows a disassembled view of an oil pump used to lubricate the cutting chain. The pump gives very little trouble.

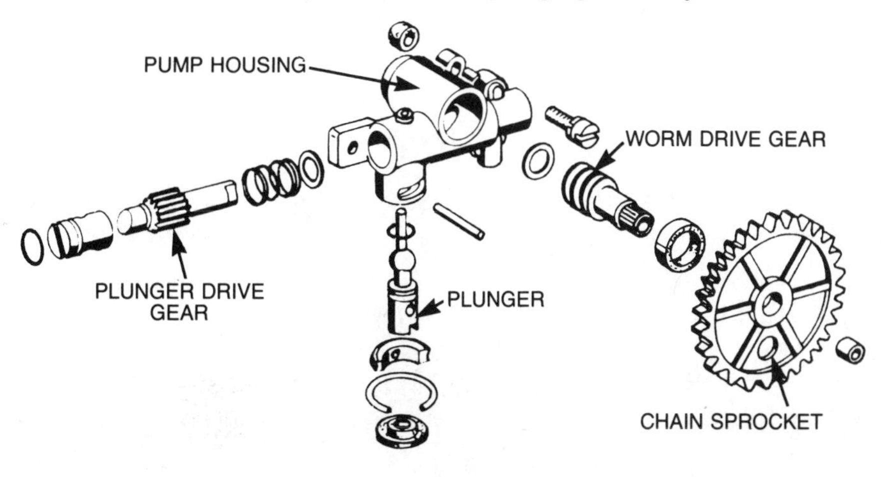

FIGURE 7-5
Disassembled chain saw oil pump
Courtesy Stihl Corporation

Before condemning it, make sure that the oil is clean and that all openings are clear. If the pump parts show signs of wear, the entire unit should be replaced. If the pump is disassembled, replace all gaskets and sealing rings.

8 FUEL SYSTEM

The fuel system stores liquid fuel and delivers it in the form of a vapor to the engine combustion chamber. Most fuel systems are made up of (1) a fuel tank to store the liquid fuel, (2) fuel lines to deliver fuel to the carburetor, (3) a fuel filter or strainer to help keep foreign matter out of the carburetor, (4) an air cleaner to clean the air as it is drawn into the carburetor, and (5) a carburetor to mix the fuel and air in the correct ratio and to deliver the required amount to the engine. Fuel pumps are required on engines that use remote fuel tanks or that have tanks mounted in such a position that gravity will not deliver the fuel to the carburetor. A fuel gauge is sometimes located on the fuel tank to register the amount of fuel in the tank.

PURPOSE OF FUEL SYSTEM

The purpose of the fuel system is to deliver to the combustion chamber an air-fuel mixture of the correct ratio, so as to provide the necessary amount of mixture for varying speed and load conditions.

Most small engines burn a mixture of gasoline and air, and the mixture must enter the engine in the form of a vapor. The ratio for normal operation is approximately one part of gasoline to 15 parts of air by weight. The amount and the ratio of the mixture entering the combustion chamber determine the speed and power that the engine delivers.

Because temperature affects the ability of gasoline to vaporize, the mixture must be "enriched"—that is, given more fuel to air—when the engine is cold. Because conditions may change within an engine, the carburetor must be adjusted to maintain an efficiently operating engine. A linkage arrangement permits the operator to control engine speed. This linkage is usually connected to a governor to maintain a more or less constant speed regardless of load.

FUEL

Most manufacturers recommend the use of a "regular" grade of gasoline. But the use of lead-free or low-lead gasoline will result in reduced combustion chamber deposits and improved engine life. Some recommend the use of a fuel with an octane rating of at least 90. Of major importance is that the fuel be clean. Do not use stale gasoline, as it tends to result in gum formations. If the engine is to stand idle during the off season, drain the fuel tank. Run the engine until the carburetor and fuel tank are dry. Doing so prevents the formation of gum in the carburetor.

FUEL TANK

A fuel tank is used to store fuel, which is normally gasoline in the four-stroke cycle engine and a mixture of gasoline and oil in the two-strike cycle engine. Most engines have the tank located above the carburetor so gravity causes fuel to flow to the carburetor. A fuel tank cap is used on the fill opening to keep fuel from splashing out and dirt from falling in. The cap must be vented to let air in above the fuel, to force fuel from the tank when the engine is running.

The fuel tank may be made of steel or plastic. Steel tanks are fabricated from pressed steel, and the joints are soldered. Water that enters a steel tank and remains for any length of time may cause rust. The resulting sediment may get past the filter and into the carburetor, where it can cause operational troubles. A steel tank can be soldered should a leak develop. With a plastic tank, however, rust is not a problem. Should a plastic tank crack, a standard epoxy cement can be used to repair cracks.

Some engines have the carburetor mounted directly on the fuel tank. The machined surface on top of the tank must be flat to provide an adequate seal between the tank and carburetor. You can check this seal by using a straight edge and a feeler gauge. A 0.002-inch feeler gauge should not enter between the tank and straight edge.

A fuel filter may be used in the tank to the fuel line adapter. The adapter may or may not employ a shut-off valve. To clean the filter, remove the adapter from the tank. Clean the filter in a cleaning solution, and blow it dry with air. Should foreign matter—such as dirt, rust, water, or other particles—accumulate in the tank, drain it and remove it from the engine. Flush it with water until it is clean, then completely dry with air. Figure 8-1 shows a typical fuel system for a two-stroke cycle engine, as used on a snowmobile.

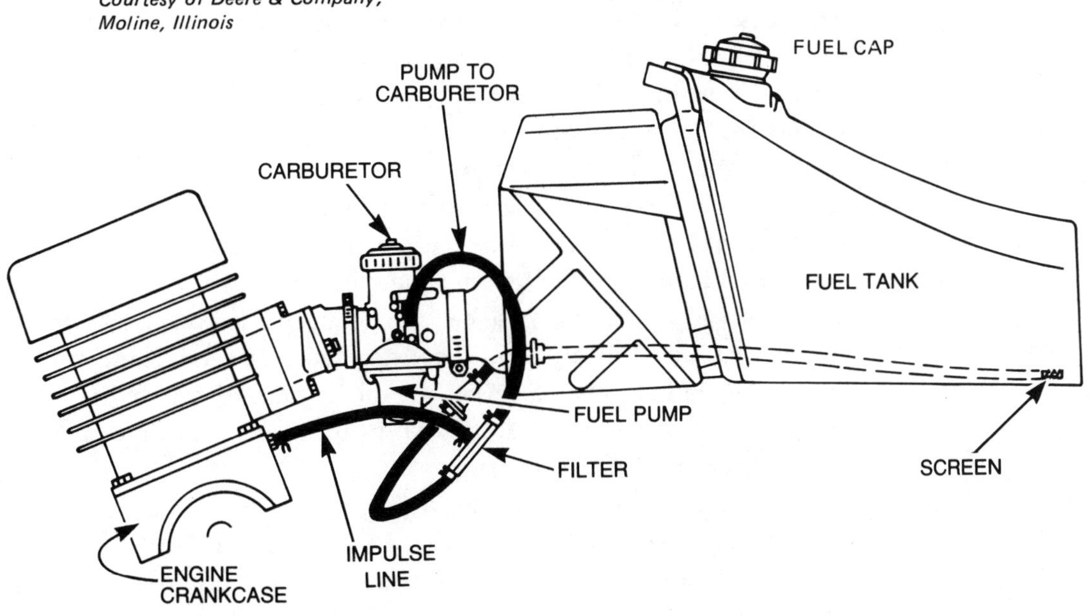

FIGURE 8-1
Two-stroke cycle engine fuel system
Courtesy of Deere & Company, Moline, Illinois

FUEL LINES

Fuel lines may be used between the fuel tank and fuel pump and between the fuel pump and carburetor. The most common type of fuel system is the gravity feed with the tank located above the carburetor. A fuel pump is unnecessary with this type of installation. Just a fuel line connects the tank to the carburetor. If the carburetor is located on top of the fuel tank, no fuel line or pump is required.

Fuel lines may be made of neoprene, steel, copper, or nylon. Neoprene is most commonly used. Check all the lines to make sure they do not leak, either through the line itself or at the connections. Make sure the line is clean and clear on the inside. Blowing air through the line—particularly a neoprene line—should clean it out. Metal lines are attached with fittings of either the flared or compression type. The neoprene line is held in place with metal clamps. Most clamps can be spread open for removal or installation with a pliers. Some may use a clamp that is held in place by a bolt or screw.

When installing new lines, make sure they have enough slack so as not to be affected by vibration. They should be routed so they do not kink, get pinched, or touch any part of the exhaust system. When installing a metal line having a flared end, you need a flaring tool to make a new flare. When installing a metal line that has a compression fitting, you must use a new compression sleeve.

FUEL FILTERS

The passages in the carburetor are small. Any foreign matter (such as dirt, lint, water, or rust) that gets into the carburetor may restrict the flow of fuel through these small openings, resulting in poor engine operation. To reduce this possibility, some type of device is used to filter the fuel before it enters the carburetor.

In most cases the filtering device takes the form of a very fine mesh bronze screen. This screen is generally located in the fuel line adapter. All fuel leaving the tank must pass through it. The mesh of the screen is fine enough to keep out everything except extremely small particles. The adapter is constructed in such a manner that the outlet and screen are raised off the bottom of the tank a little bit, so that foreign matter settles to the bottom of the tank where it is not drawn into the carburetor.

When the carburetor is installed on top of the fuel tank, a filter (screen) and check-ball may be located at the bottom of the pick-up pipe or standpipe assembly. In either installation, the screen is cleaned by soaking it in a cleaning solvent and blowing it dry with compressed air.

Some engines use a cartridge-type filter located in the fuel supply line. When fuel no longer flows freely through the filter, the entire unit is replaced. The flow is checked by removing the line from the output side of the filter. The fuel should flow through the filter without restriction.

Below the tank on some larger horizontal-shaft engines may be mounted a regular fuel filter, which utilizes a filter screen and sediment bowl (Figure 8-2). A shut-off valve is usually incorporated into the assembly. To service the bowl, should dirt and/or water get into it, close the shut-off valve and disassemble the filter. Clean the screen by soaking it in a cleaning solution and blowing it dry. Clean out the sediment bowl and wash it in gasoline. Check for chipped edges on the glass and for warped sealing edges on the body. Use a new gasket when assembling the unit.

A fine mesh screen may be located at the carburetor inlet opening. Remove this by disconnecting the fuel line and unscrewing the carburetor inlet fitting. Clean the screen in a cleaning solution.

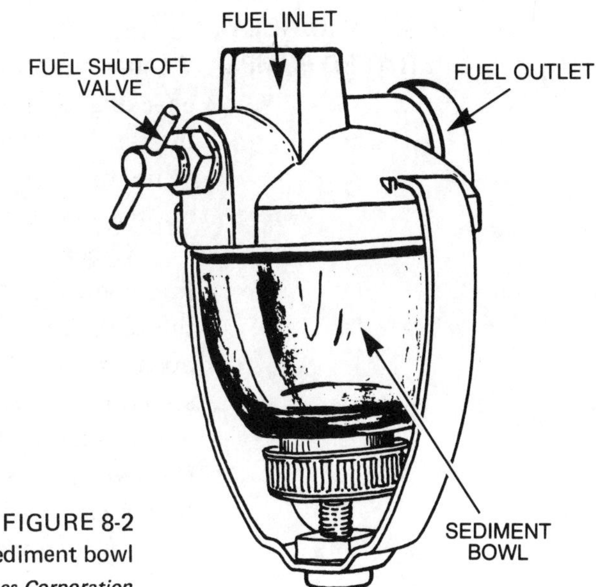

FIGURE 8-2
Fuel filter with sediment bowl
Courtesy Clinton Engines Corporation

FUEL PUMPS

When an engine has a remote fuel tank, or when the tank is mounted in such a position that gravity cannot deliver fuel to the carburetor, then a fuel pump is necessary. The pump draws fuel from the fuel tank and delivers it to the carburetor according to the carburetor's demands. The fuel pump may be mechanically or vacuum operated. The mechanical pump is actuated by the camshaft or crankshaft, while the vacuum-operated pump is actuated by the pulsating negative pressure (vacuum) with the crankcase.

Most two-stroke cycle engines that require fuel pumps use the vacuum diaphragm. The pump diaphragm is actuated by the crankcase's vacuum pulsations. Many of these pumps are integral parts of diaphragm-type carburetors.

CHECKING FUEL PUMP OPERATION

A fuel pump normally gives dependable service over a long period. It performs but one function and that is to supply sufficient fuel to the carburetor. As long as the engine is getting enough fuel to operate properly under all load and speed conditions, and as long as you see no external fuel leaks, do not disturb the pump. If the engine is being starved—that is, if it is short on fuel, as indicated by hard starting, a lack of power, or sputtering when accelerating—first check the tank to see if you have enough fuel. Then check the filter and screen to make sure they are clean. Also make sure that the fuel lines are open and that they do not leak. Be sure the carburetor is adjusted correctly.

MECHANICALLY OPERATED PUMPS

When a line runs from the outlet of the fuel pump to the carburetor, checking to find out if fuel is reaching the carburetor is a simple matter. Disconnect the fuel line at the carburetor inlet or pump line. If no fuel comes out, or if there is just a trickle, then either fuel is not reaching the pump or the pump is not functioning.

Check to make sure fuel is reaching the pump. Disconnect the fuel line from the tank at the fuel pump. If the pump is lower than the tank, a good stream of fuel should flow if the line and screen are open. If not, check for fuel in the tank and then whether or not the screen and line are open. If the tank is lower than the pump, blow air into the tank. If the line and screen are clear, a stream of fuel should flow when air is applied to the tank. If not, check the line and screen. If fuel can reach the pump but not the carburetor, then the pump is at fault. Remove the pump and either replace or rebuild it.

VACUUM-OPERATED PUMPS

Positively checking the operation of the vacuum-operated diaphragm pump is more difficult when it is an integral part of the carburetor. On the two-stroke cycle engine, make sure that the correct gas-oil mixture is being used, that the oil level in the air filter is not too high, and that the air filter is clean. Make sure the exhaust system is not clogged. Check to make sure the crankcase oil seal around the crankshaft is not leaking. Leakage is indicated by a collection of dirt and oil on the outside of the engine around the seals.

If a line from the fuel tank to the fuel-pump/carburetor combination is used, remove the line at the fuel pump and make sure fuel flows freely. With the line disconnected, place your finger over the fuel inlet connection while cranking the engine. You should feel a suction on your finger if the pump is working.

Here is another method for checking whether a pump-and-diaphragm carburetor is delivering fuel. Tightly close the choke valve and crank the engine several revolutions. Then remove and inspect the spark plug. If fuel is being delivered, the plug will be wet. If it is completely dry, no fuel is reaching the combustion chamber. The lack of fuel mixture in the combustion chamber could also be the result of no compression or a leaking reed or rotary valve.

Another check, if the plug is dry, is to remove the spark plug and pour a small quantity of fuel into the cylinder. Replace the plug and try to start the engine. If the engine fires, then fuel is just not getting to the combustion chamber.

A leak tester, which positively determines whether a leak is present, consists of a pressure bulb that is used to apply air pressure at the fuel inlet opening. The carburetor and fuel pump assembly should hold about 5 pounds of pressure for at least one minute without dropping. If the pressure does not hold, check the cover screws for tightness. If the screws are tight, there is an internal carburetor and/or fuel pump leak. The pump and carburetor must be removed and disassembled for service.

REPAIR FUEL PUMPS

To service the pump, you must remove it from the engine. If the entire unit requires service, then the combination fuel pump and carburetor must be removed.

Before disassembling the fuel pump, find out what is available in the way of repair kits. Some vacuum-operated pumps are non-serviceable, so the entire unit must be replaced. Others that are integrated with the carburetor can be rebuilt; repair kits that contain gaskets, valves, and diaphragms are usually available. It is generally advisable to replace the carburetor diaphragms when replacing the pump diaphragms.

Disassembly and assembly are obvious procedures. Before disassembly mark corresponding surfaces if they must go back together in the same relative positions. A file mark can be made on the edges. Figure 8-3 shows a diaphragm-type vacuum-operated pump attached directly to the carburetor. When assembling, tighten all screws evenly to prevent warping and to assure a complete seal. Always keep in mind that the objective is to have a vacuum-tight unit. One side of the diaphragm is the wet side, the other is dry. Make sure all surfaces are clean and fit properly.

To service the mechanically operated diaphragm pump, disconnect the fuel lines and remove the mounting screws that hold the pump to the engine (Figure 8-4). Clean the outside of the pump. Some pumps cannot be disassembled, so the complete unit must be replaced. Using a file, make an indicating mark across the pump body

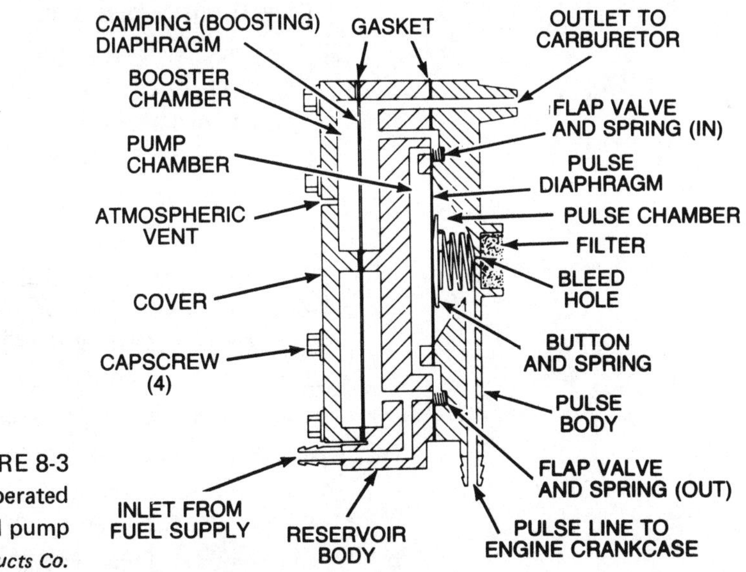

FIGURE 8-3
Diaphragm-type vacuum-operated fuel pump
Courtesy Tecumseh Products Co.

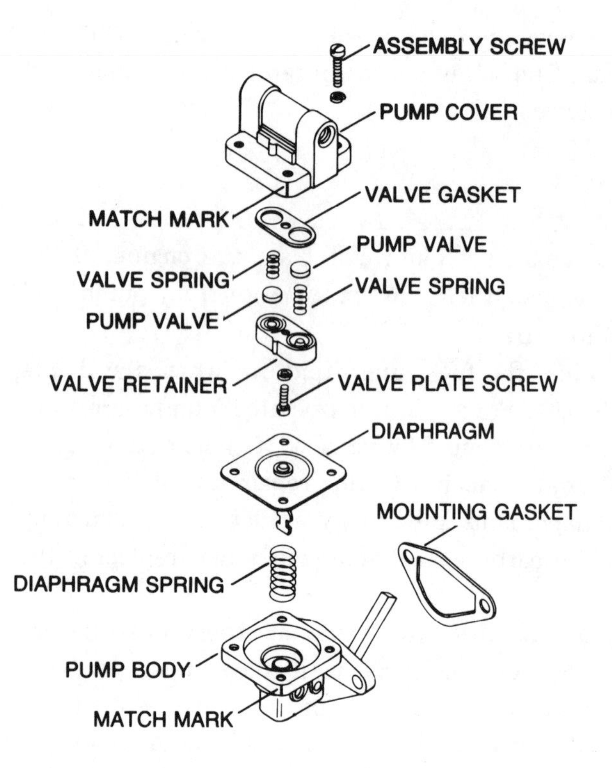

FIGURE 8-4
Mechanical-type diaphragm fuel pump (Disassembled)
Courtesy Kohler Co.

and lower section so they can be reinstalled in the same relative positions. Remove the screws and pull off the cover. Replace the diaphragm, gaskets, and valves. You may have to remove the diaphragm drive lever pin to remove the diaphragm. On some pumps the diaphragms can be unhooked from the pump levers by holding the lever and pushing down on the diaphragm assembly. Make certain all surfaces are clean and fit properly. Tighten the cover screws evenly to be assured of a tight fit. Operate the pump by hand before installing it on the engine to make sure it operates. There should be suction on the inlet side and pressure on the outlet side.

When installing the pump, be sure the pump lever makes contact with the operating cam on the camshaft or crankshaft. Heavy grease can be used to hold the lever in place during installation if the lever needs to be held in an upright position. Revolve the crankshaft to make sure the lever is correctly positioned before tightening the attaching bolts. Reconnect the fuel lines.

AIR CLEANER

To provide for maximum engine protection, the air cleaner should filter the air completely free of dirt and dust before it enters the carburetor. Dirt inducted through the carburetor due to an improperly serviced or inadequate air cleaner element will wear out an engine faster than long hours of operation. A clogged air cleaner can result in too rich an air-fuel mixture, which may in turn lead to the formation of harmful sludge. While the filter must be of a very fine mesh to keep out the dirt, it will prevent the free flow of air if the material is too fine or if it becomes clogged. Again, the result is too rich an air-fuel mixture.

The air cleaner should be serviced frequently to prevent clogging and to prevent dust and dirt from entering the engine. A properly serviced air cleaner protects the internal engine parts from dust particles in the air. The interval between servicing varies with use. Follow the manufacturers recommendations, found in the operator's manual. The time intervals between servicing vary from 10 to 25 hours. Under extremely dusty operating conditions, the cleaner may have to be serviced every time the engine is used. The cleaner should be inspected regularly to make sure it is not plugged with dirt. Clean it as soon as there is any accumulation of dirt. The air cleaner can usually be removed by simply unscrewing a wing nut or one screw. After its removal from the engine, the assembly can generally be pulled apart. Do not tip the oil bath cleaner, because the oil will spill out.

TYPES OF AIR CLEANERS

The numerous types of air cleaners are all designed to perform the same function.

The foam (polyurethene) cleaner is the most commonly used type on small engines. The foam cleaner is like a very fine sponge with very small openings. When oiled, it traps a lot of dirt (Figure 8-5). It can be shaped to fit variously sized containers. Some cleaners use a foam sleeve over a dry cleaner as a precleaning element, to increase the filtering capacity. The oil foam filter becomes clogged with dirt over a period of time. A screen or container is always used to hold the foam in place. The element can be washed and re-oiled when it becomes dirty.

A dry paper filter is found on some engines (Figure 8-6). The fine-grained porous paper serves to prevent dirt from entering the carburetor. Some paper filters are treated so the dirt and dust are trapped on the paper.

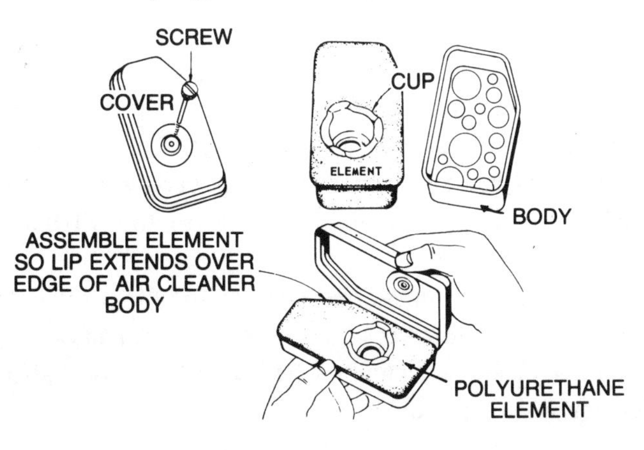

FIGURE 8-5
Polyurethane-type air cleaner
Courtesy Briggs & Stratton Corp.

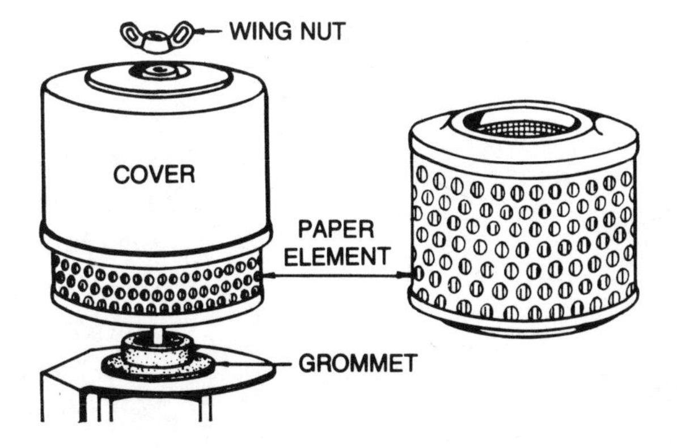

FIGURE 8-6
Paper-type air cleaner element
Courtesy Briggs & Stratton Corp.

A metallic mesh or screen-type cleaner, which is used on some engines, is serviceable. When clean it has less resistance to the flow of air than some other types of cleaners, but it may not be as effective.

The oil bath filter, sometimes referred to as a "heavy duty filter," uses oil to trap the dirt. Although it requires less service than the other types of filters, the oil level should be checked regularly. A filter element of metallic mesh is used above the oil. As air entering the cleaner passes over the surface of the oil, it picks up some of the oil and carries it through the filter element. The oil saturates the filter material and drops back down into the oil reservoir. As it does so the particles of dirt in the filter material are washed down by the oil bath. A lot of dirt can accumulate in the bottom of the reservoir before servicing is necessary. Figure 8-7 shows a typical oil bath air cleaner disassembled.

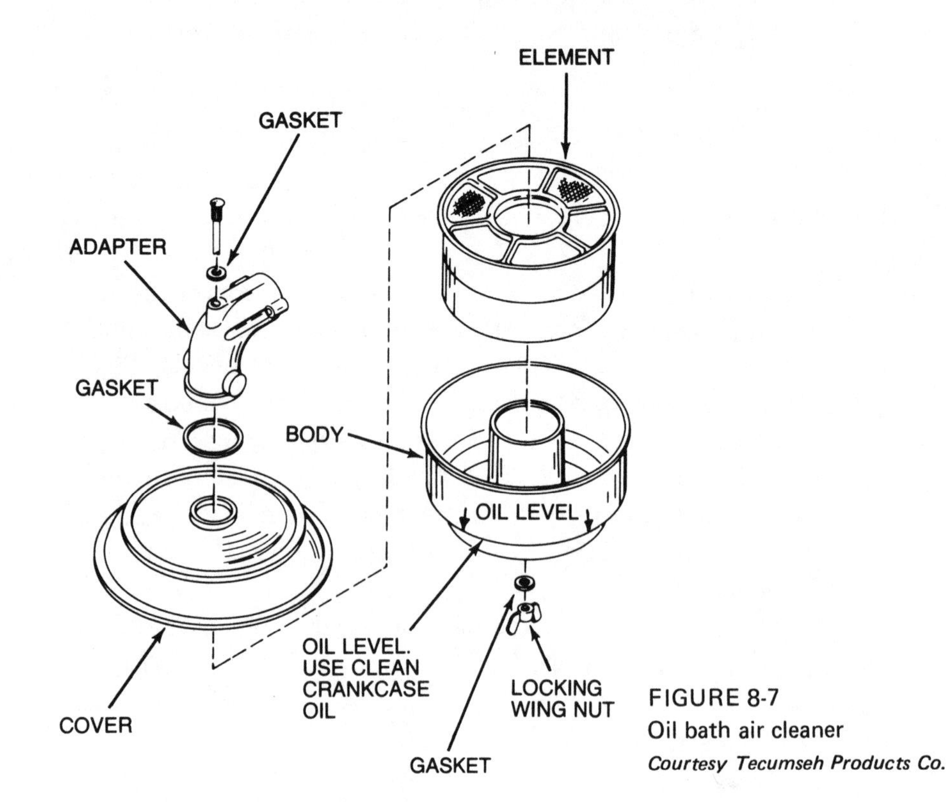

FIGURE 8-7
Oil bath air cleaner
Courtesy Tecumseh Products Co.

SERVICING AIR CLEANERS

The air cleaner (filter) is usually attached to the carburetor air horn by a bolt and wing nut or by a screw. Removing the wing nut or screw permits you to remove the cleaner from the carburetor. The cleaner comes apart after being removed. Do not tip the oil bath type filter, or the oil may spill out.

After disassembly, clean the cover and base or body in a solvent such as mineral spirits or a commercial cleaner. You can also use some liquid household cleaners, which when mixed with water dissolve grease, to clean engine parts.

The foam (polyurethane) type of cleaner elements, when dirty, should be washed in a mixture of liquid detergent and water. After washing, squeeze dry. When the element is dry, saturate it with engine oil and let the surplus oil drain away before installing. Replace the element if it has disintegrated or if it has holes so that it cannot filter properly.

When a dry paper element is used, gently tap the element on a hard flat surface to knock out the dirt. Air can be used to blow out dust and dirt. Holding the element up to a light shows if the element is dirty. If the paper is treated, somewhat oily, it must be replaced when dirty. If the paper is not treated, wash the element in soap and water, rinse out the soap, and let dry. Replace the element if it is perforated, torn, or plugged with dirt. Do not oil this type of filter.

A few engines use a metallic mesh cleaner. The mesh can be washed in a solvent and blown dry with air. If the mesh does not completely fill the retainer screen, the element should be replaced.

When an oil bath filter is used, the oil should be drained and the complete unit cleaned if the oil is dirty. Fill it with clean oil to the proper level, which is marked on the filter bowl. If the oil is clean and at the proper level, merely clean the filter unit. If the oil is clean but the level is low, add oil to bring the level up to the mark. Regular engine oil is used in the filter.

Replace the air cleaner gaskets and mounting gaskets that are worn or damaged. Doing so prevents dirt from entering around the air cleaner.

CARBURETORS

The purpose of the carburetor is to provide the correct air-fuel mixture to the engine for varying speed and load conditions. The carburetor must mix air and fuel into a vapor that is correctly proportioned so as to have a combustible mixture that meets the various operating conditions.

CARBURETOR OPERATION

Air is drawn in through the carburetor air horn in such a manner that it picks up fuel. The suction that draws the air through the carburetor is created by the intake stroke of the piston. As the piston moves downward a partial vacuum is created in the cylinder because the pressure within the cylinder is lower than atmospheric pressure. With the intake valve or port open, a passageway through the cylinder is connected to the carburetor. Air can then flow through the carburetor into the cylinder. The amount of air flow is controlled by a disc valve known as the "throttle valve" or "plate," located in the carburetor throat. The opening or closing the throttle valve regulates the amount of air drawn into the engine. Figure 8-8 is a sectional view of a float-type carburetor .

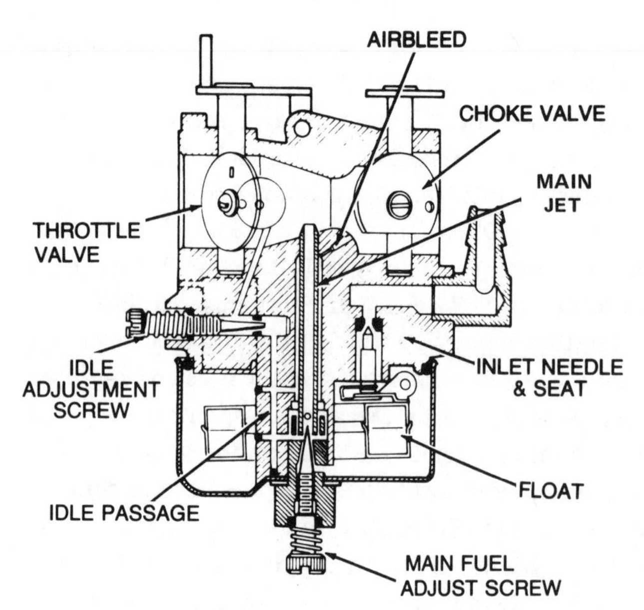

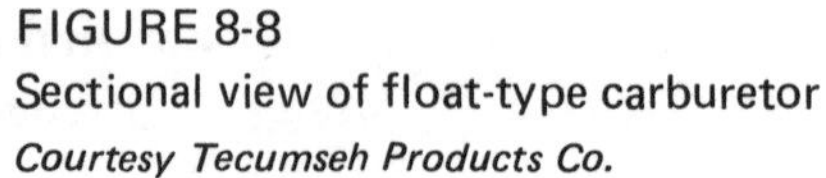
FIGURE 8-8
Sectional view of float-type carburetor
Courtesy Tecumseh Products Co.

Gasoline from the float bowl or fuel chamber is present in a metered fuel passage, which protrudes into the carburetor air passageway. This air passage, which leads from the air horn to the cylinder, has a constriction called the "venturi." The venturi is simply the narrowing of the passage to create an increase in a vacuum as the air moves past this point. The metered fuel passage (main jet) extends into the venturi in such a manner that the vacuum created by the flow of air draws fuel from the main jet. As the fuel is drawn from the jet, it is mixed with air and forms a vapor. The normal air-to-fuel mixture is approximately 15 parts of air to one part of gasoline by weight. The size of the opening in the metering jet determines the amount of fuel flow.

On most engines you have to be able to vary the speed. So the carburetor has to be able to accommodate variable conditions. Doing so is the purpose of the throttle.

When an engine is cold the fuel does not vaporize as readily as when warm. So you must be able to enrich the mixture when the engine is cold. A choke valve, inserted in the air horn, can be used to restrict the flow of air. It thus gives the mixture more fuel in proportion to air, by causing more gasoline to be drawn out of the main metering jet.

At low speeds the flow of air through the carburetor is restricted by the throttle valve. The reduced flow of air cannot draw enough fuel from the main metering jet to keep the engine running properly. To compensate for this restriction, an idle passage, located in the carburetor barrel above the throttle valve, provides additional fuel during periods of idle. An adjustable needle valve is inserted into the passageway to permit the idle mixture to be enriched or "leaned out" for more efficient low-speed operation. An adjustable needle valve, in the main metering jet, also permits changing the amount of gasoline that enters the air stream. A correct mixture of air and fuel makes for a more efficiently operating engine.

TYPES OF CARBURETORS

Carburetor design is affected by engine design. Some engines require a side draft carburetor, which is attached to the side of the engine directly onto the intake port leading to the combustion chamber. Other engine designs require an updraft carburetor, which is mounted below the intake port so the fuel mixture must be lifted up through the intake manifold and into the combustion chamber.

All the different models and designs of carburetors basically operate in one of two ways: (1) as float feed systems or (2) as diaphragm systems.

1. In the *float feed carburetor*, fuel must be maintained in the float bowl at a specified level for the engine to get the correct amount of fuel. So the carburetor and engine must remain in a relatively level position.

2. Most *diaphragm* carburetors are constructed so they can operate in any position. This type of carburetor must be used on engines that power chain saws, grass cutters, and other machines that wind up in various positions. Figure 8-9 is a schematic of the diaphragm-type of carburetor.

While sizes, shapes, and appearances differ among the various carburetor installations, all carburetors perform the same function and for the most part consist of the same basic components. With

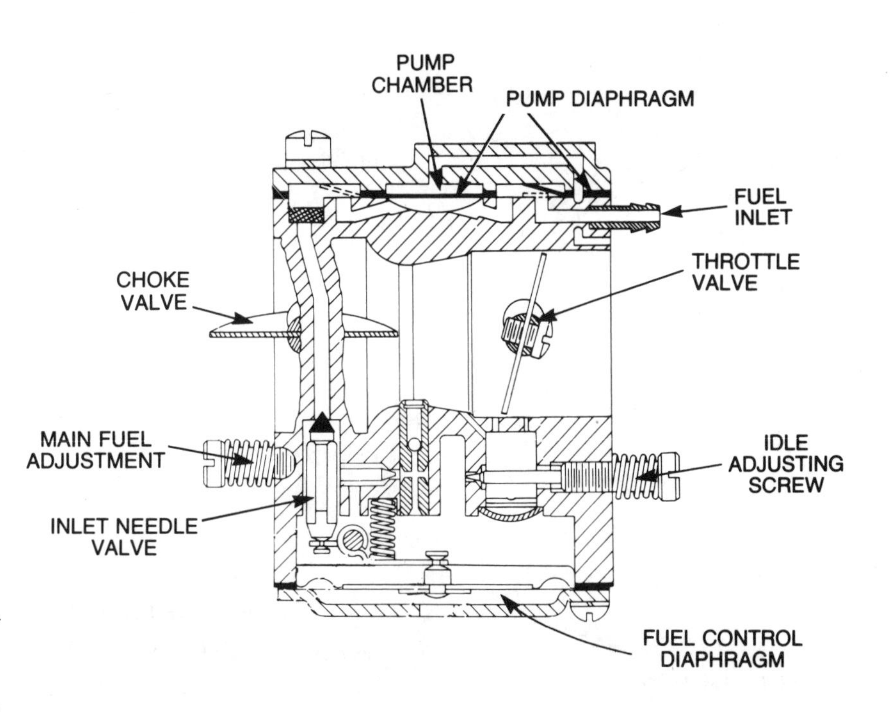

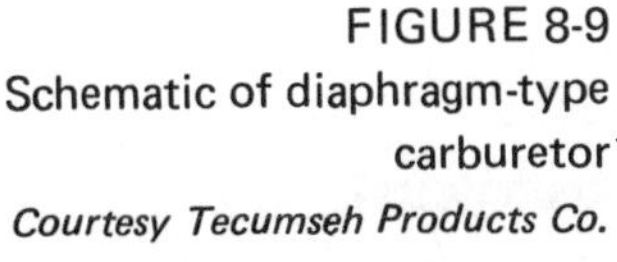
FIGURE 8-9
Schematic of diaphragm-type carburetor
Courtesy Tecumseh Products Co.

few exceptions all carburetors have the following external adjustments: (1) an idle mixture adjustment, (2) an idle speed adjustment, and (3) a main fuel mixture (high-speed) adjustment (Figure 8-10). The float level is adjustable on a float-type carburetor, but once the setting is correct there should be no reason to change it.

CARBURETOR TROUBLE SHOOTING

If the engine is functioning properly, there is no reason to disturb the carburetor. The carburetor has but one function to perform, and that is to mix fuel with air and deliver it to the combustion chamber for all speed and load conditions. A carburetor can malfunction basically only three ways: (1) Its mixture may be too lean; (2) it may be too rich; or (3) the carburetor may leak. The operation of the carburetor may also be affected by foreign matter, solid or liquid, that retards the flow of air or the movement of fuel and/or air.

Always check to make sure the trouble exists in the carburetor before removing it. Unless foreign matter gets into the carburetor, about the only things that should need to be done would be to readjust the idle speed, idle mixture, or fuel mixture. Some float carburetors have a bowl drain valve on the bottom of the float bowl. Press the drain valve occasionally before servicing, and let a small amount of fuel flow out. Catch the fuel in a container. If water and/or dirt comes out, you may safely assume that water or dirt is in the system. The adjustments and bowl drain are shown on the carburetor in Figure 8-10 (a commonly used side mounted carburetor).

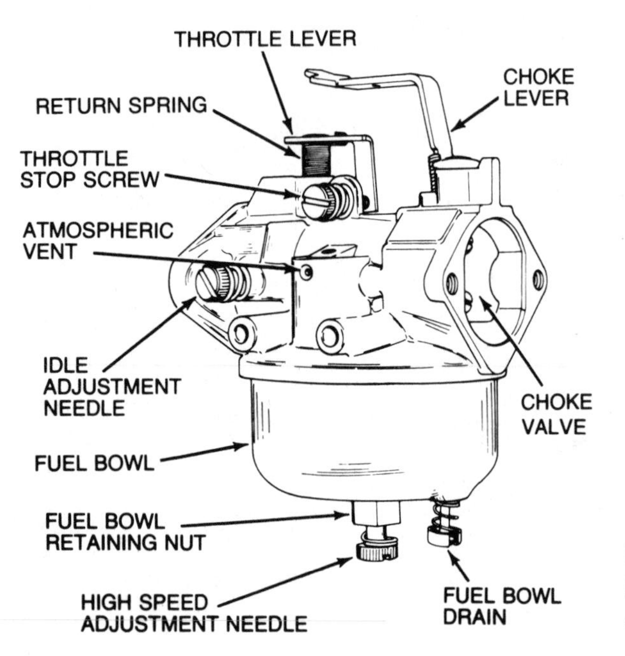

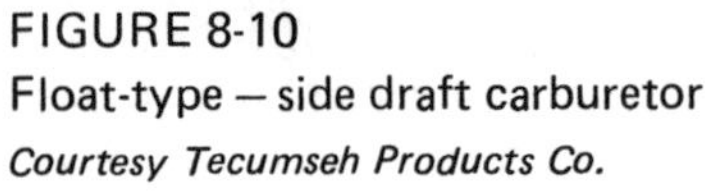
FIGURE 8-10
Float-type — side draft carburetor
Courtesy Tecumseh Products Co.

Certain conditions may indicate faulty carburetion on a running engine:

1. Black sooty exhaust smoke and sluggish operation may indicate too rich a mixture. Readjust the main fuel mixture adjusting screw to lean out the fuel flow.

2. If the engine misses and backfires at high speed, it indicates too lean a mixture. Readjust the main fuel mixture adjusting screw to permit more fuel to enter the airstream.

3. If the engine, when starting cold, starts up and then dies, then the mixture is too lean. Open the main fuel mixture adjusting a screw a small amount.

4. When the engine runs roughly or stalls at idle, then the idle speed is set too low or the idle mixture needs to be readjusted.

5. If the engine fails to start—and assuming the ignition system is functioning properly, as indicated by a good spark at the spark plug—check to see that there is fuel in the tank. Make sure that the tank shut-off valve is open and that fuel is reaching the carburetor. Check to make sure the choke opens and closes properly. Make certain the air cleaner is not clogged.

Check the carburetor adjustments on the chance that they may have been inadvertently turned. Always adjust them to specifications. Lacking the manufacturer's specifications, turn the main fuel adjustment in until it is tight. Do not force it, or it may groove the needle valve. Back out approximately 1½ turns. Do the same for the idle adjustment, only turn it out approximately one turn. These adjustments should enable the engine to start and run.

6. If you fail to get the proper reactions when making adjustments on a running engine, there are internal carburetor problems—probably foreign matter in some of the passages or an air leak.

CARBURETOR REPAIR

Before removing the carburetor, check the choke linkage arrangement—how it opens and closes—so the linkage can be installed in the proper manner. When the choke is manually applied, the choke valve must close tightly. When the choke is off, the valve must be wide open. The same applies when an automatic choke is used.

Check the throttle linkage as well as the governor linkage. Mark it if necessary, or make a sketch of it, so it can be installed in exactly the same location and position from which it was removed. Do not bend or distort the linkage in any way. This is very important for correct throttle and governor action. Figure 8-11 illustrates a common linkage arrangement for a manual throttle and governor control.

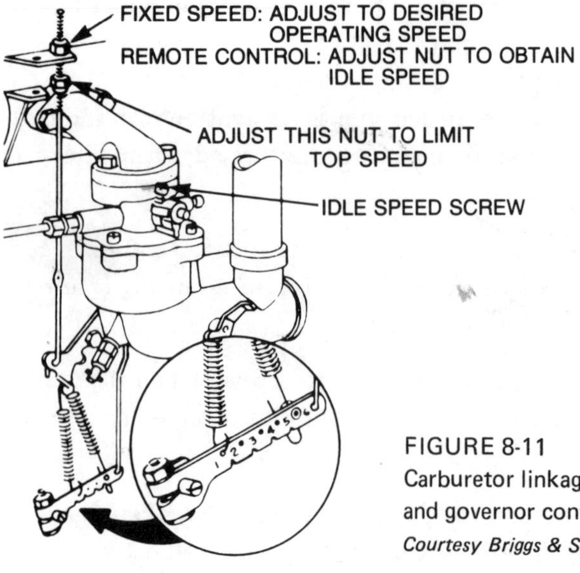

FIXED SPEED: ADJUST TO DESIRED OPERATING SPEED
REMOTE CONTROL: ADJUST NUT TO OBTAIN IDLE SPEED
ADJUST THIS NUT TO LIMIT TOP SPEED
IDLE SPEED SCREW

FIGURE 8-11
Carburetor linkage – throttle and governor control
Courtesy Briggs & Stratton Corp.

Carburetor difficulties usually originate with improper adjustments or with dirt, gum, or varnish on components. A gum or varnish-like substance is generally the residue of evaporated fuel. Fuel left in the carburetor or sediment bowl for a period of time evaporates, leaving a gummy varnish-like coating or residue.

If a carburetor is to be disassembled always use new gaskets when reassembling it. All metal parts should be cleaned in alcohol, acetone solvent, or a special carburetor cleaner. Do not clean any diaphragms, nylon, or other nonmetallic parts in the solvent. Cleanliness is essential in carburetor repair.

Repair kits are available for most carburetors. These kits usually include gaskets, a float pin (for float-type carburetors), a float inlet needle, and a seat. If a new float needle valve and seat are installed because the needle valve shows signs of wear, the float level must be

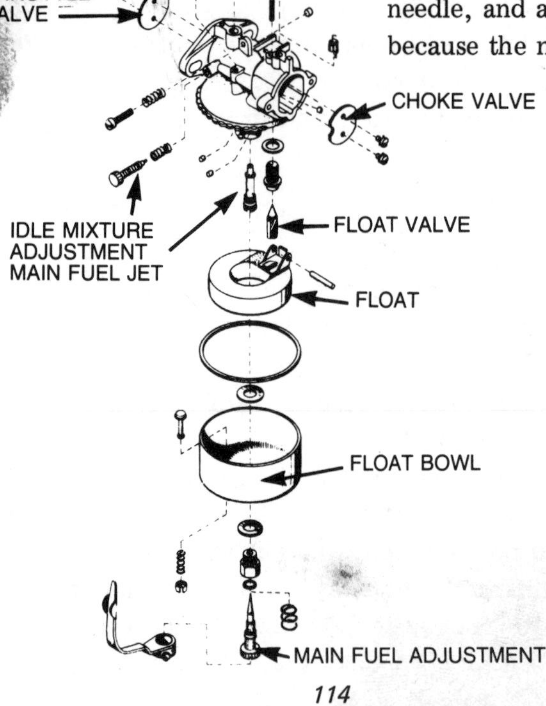

FIGURE 8-12
Disassembled view of float-type carburetor
Courtesy Clinton Engines Corporation

checked. The needle valve seat cannot be removed in some carburetors. The level is changed by bending the lip on the float. Check the manufacturer's specifications for the correct float level setting. If either the throttle valve shaft or the carburetor body where the shaft goes through is worn, replace the carburetor. Figure 8-12 shows a disassembled float-type carburetor.

In most cases with the float-type carburetor, servicing involves cleaning all the parts, blowing out all the passages, and installing new gaskets. Always remember to set the adjustment screws before trying to start the engine.

If you have to disassemble the diaphragm-type carburetor, replace the diaphragms, since they may tear or distort upon disassembly. The same advice applies to the combination vacuum diaphragm fuel pump and carburetor. Make sure replacement parts are available. This carburetor depends on vacuum for its operation, so any air or vacuum leak interferes with its functioning properly. Figure 8-13 is a disassembled picture of a diaphragm-type carburetor.

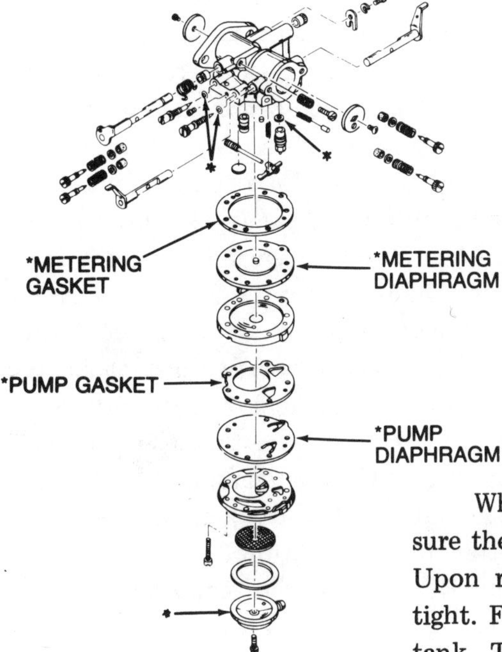

FIGURE 8-13
Disassembled diaphragm-type carburetor
Courtesy Tecumseh Products Co.

When the carbureator is installed on top of the fuel tank, make sure the tank surface where the carburetor fits is clean and smooth. Upon reassembly, always install a new gasket, which must be airtight. Figure 8-14 shows a carburetor installed on top of the gasoline tank. The fuel pickup tube extends into the tank. The parts are disassembled and labeled for identification.

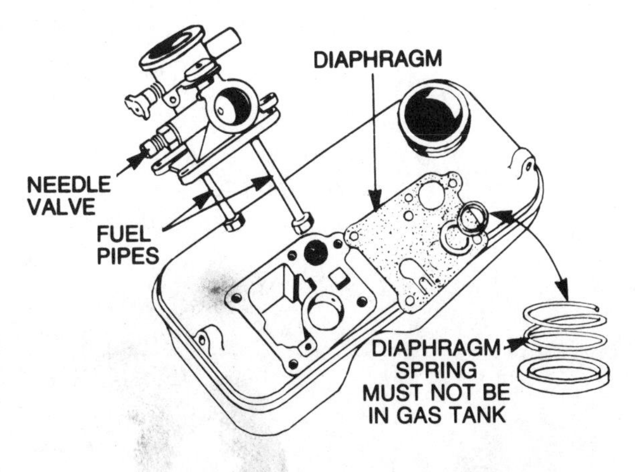

FIGURE 8-14
Exploded view of tank top carburetor
Courtesy Briggs & Stratton Corp.

Some carburetors are equipped with an automatic choke. Before disassembly check the exact linkage setting, so as to reinstall it in the exact location. If the vacuum diaphragm, which operates the choke, is punctured or stretched, replace it.

In most cases, when servicing a carburetor, the rule is: the fewer things disturbed, the better its chances of good operation. Thoroughly cleaning all the parts and blowing out all the passages generally takes care of most problems. Always make sure the problem is in the carburetor before removing it from the engine for disassembly. If a carburetor is kept clean, it will operate for a long time without service other than adjustments.

CARBURETOR ADJUSTMENTS

Normally three adjustments can be made to most carburetors: (1) the idle speed, (2) the idle mixture, and (3) the main fuel (high-speed) adjustment. Figures 8-9, 8-10, 8-12, and 8-14 show the adjustments on different carburetors. The engine should be running and at normal operating temperature when you make these adjustments.

If an improper carburetor mixture is indicated, stop the engine and turn both the main fuel adjustment screw and the idle mixture screw all the way in until they bottom lightly. Do not force them because you may damage the needle valve. For a preliminary setting, turn the main fuel adjustment out approximately 1½ turns and the idle adjustment out about one turn. Start the engine and run it under load (that is, actually operating the equipment, if possible), turn the main fuel adjustment in until the engine slows down (too lean), then turn the screw out until the engine begins to slow down again (too rich). Turn it back halfway between too rich and too lean. Adjust the idle mixture in the same way, only with the throttle set so the engine just keeps running. With the engine idling as smoothly as possible, increase the idle speed a small amount. This extra amount generally causes the engine to run better. Idle speed without load is approximately 1,200 rpm. The engine should accelerate smoothly from idle speed; if not, then the idle speed may be set too low.

Test the engine by running it under a normal load. The engine should respond to load "pickup" immediately when the throttle is opened. If the engine tends to "die," the mixture is too lean. An engine that runs roughly under load is probably too rich.

CARBURETOR CONTROLS

With hand-operated choke, the cable must be installed so that in the closed position the choke valve is tightly closed and in the open position it is wide open. The same rule applies to the automatic choke: Upon start-up of a cold engine, the choke valve should be tightly closed; when the engine is hot it must be wide open. Make sure the automatic choke linkage does not bind. If the choke fails to operate, check the vacuum hose to the choke diaphragm, if one is used. It must be tight and have no leaks. If the choke fails to open or close tightly, a linkage adjustment may take care of the problem.

The manual throttle linkage must operate freely completely opening and closing the throttle valve. The linkage can usually be set by moving the cable in or out to obtain the proper travel. Once correctly installed, the choke or throttle linkage should not change. You can now see why, should the linkage have to be removed, you must know exactly where it was originally located.

All springs must be hooked up properly for the governor to operate. Again, you can see the importance of replacing the linkage, controls, and springs in exactly the order from which they were removed.

The throttle position controls engine speed. Most engines use a remote control throttle system so the operator has complete control over the engine speed. In most installations a governor is used to prevent overspeeding. Three basic types of throttle controls are in general use:

1. One system, the remote governor control arrangement, regulates engine speed by changing the governor spring tension.

2. Another type, the remote throttle control system, operates the throttle valve directly, just like the accelerator on an automobile, while the governor prevents engine overspeeding.

3. A combination choke, throttle, and ignition shut-off remote control system may be found on a number of engines. On this system moving the remote control lever in one direction—to the start position—closes the choke valve. As the control lever is moved from choke to the fast position, the throttle valve is opened wide. Moving the lever farther causes the throttle valve to begin to close, and the engine slows down to idle. Beyond slow, the throttle valve is closed completely. Moving the lever farther to the stop position shorts out the ignition on some installations.

Unless parts have to be removed, or unless the cable clamp or cable retainer screw has loosened, the carburetor control adjustment should not change. If you carefully examine the control arrangement and watch the movement as the control is moved, you should easily see and understand what should be happening. Loosening the cable clamp usually permits changing the control movement.

9
GOVERNORS
OPERATION OF AIR VANE GOVERNORS
OPERATION OF MECHANICAL GOVERNORS
GOVERNOR SERVICE

Most small engines are equipped with a governor, whose basic purpose is to maintain, within limits, a desired engine speed even though the load may vary. The governor keeps the engine from overspeeding and controls the speed from no load to full load.

The governor normally does not require service. But the linkage or springs may wear, become distorted, or have to be disconnected when servicing the carburetor or other components. Any of these conditions may require the replacement of parts and/or adjusting for better operation. If you understand how the governor is supposed to function, you should be able to perform the needed service work.

There are two types of governors: (1) the pneumatic air vane and (2) the mechanical flyball type.

OPERATION OF AIR VANE GOVERNORS

The air vane governor is operated by air flow created by fins on the flywheel. The air vane is located inside the blower housing and linked to the carburetor throttle lever through a spring. The air vane is a flat piece of metal that is mounted and pivots on a pin. Air flow created by the flywheel causes the vane to move on the pivot pin against spring tension. The governor spring tends to open the throttle.

Figure 9-1 shows a typical air vane governor. As engine speed increases, the air flow against the vane increases and tends to close the throttle. Closing the throttle causes the engine to slow down. The engine speed at which these two forces, air flow and spring tension, balance results in what is termed "governed speed." A remote control lever and linkage permit the operator to vary the engine speed by changing the governor spring tension. As the governor regulates fuel flow (throttle opening), the engine maintains a constant speed regardless of its load.

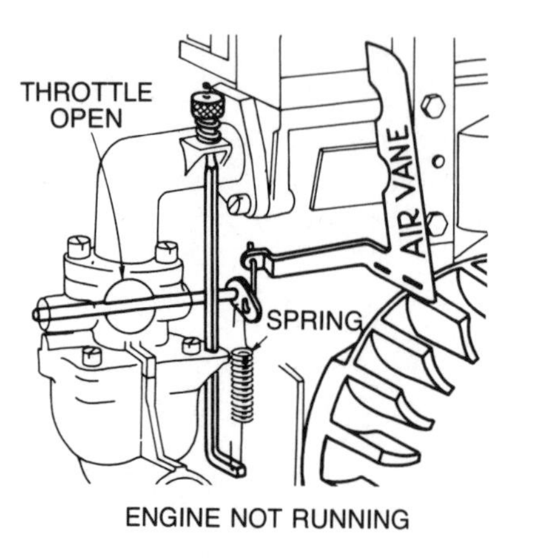

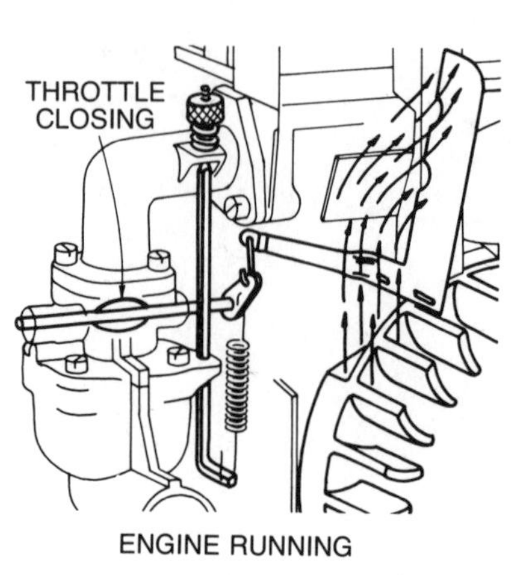

FIGURE 9-1
Air vane governor
Courtesy Briggs & Stratton Corp.

OPERATION OF MECHANICAL GOVERNORS

The mechanical governor consists of a pair of flyweights or flyballs mounted on pivots within the crankcase and turned by a gear that is in mesh with a gear on the crankshaft or camshaft. Two weights are used, and one end of each is mounted on a pivot pin. As the assembly is turned the free end of the weight moves outward due to centrifugal force. The inner ends of the weights are mounted on a sleeve or yoke arrangement, which moves in or out as the outer ends of the weights move. The movement of the yoke rotates the governor control shaft (Figure 9-2).

When the engine is not running, the flyweights come together, allowing the yoke to move toward the gear. This reaction causes the control shaft to turn and, through a spring and linkage arrangement, opens the throttle valve. When the engine is started, centrifugal force causes the flyweights to move outward, thus moving the yoke and turning the shaft, which tends to close the throttle valve against spring tension. The spring between the governor linkage and carburetor throttle tends to keep the throttle open. The engine speed at which these two forces balance is known as the "governed speed," which can be varied by changing governor spring tension. The control

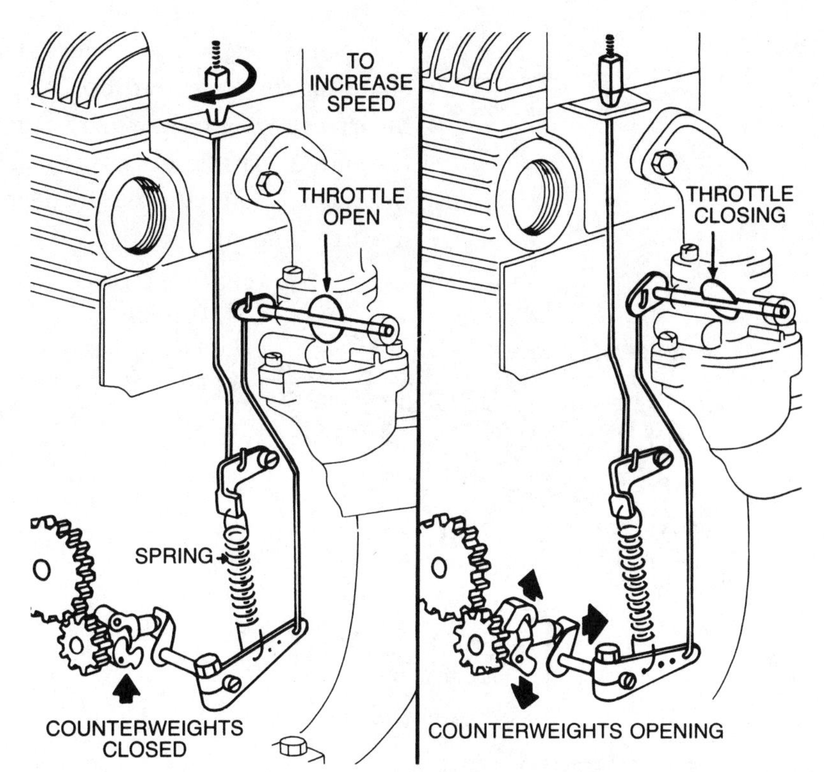

FIGURE 9-2
Mechanical-type governor
Courtesy Briggs & Stratton Corp.

is generally in the form of a lever located within the reach of the operator.

If internal problems occur with the mechanical governor the unit must be removed for service. On some engines the complete unit can be removed by unbolting it from the outside of the engine. If not, the engine must be disassembled to get at the operating mechanism.

GOVERNOR SERVICE

The governor should require very little service other than the replacement of a worn linkage or of a distorted or broken spring. Governor action can be checked by operating the engine at a uniform speed and then applying a load to the engine. Engine speed should momentarily drop off and then resume at the previous speed, assuming the engine is capable of handling the load. Watch the movement of the linkage; there should be no free play. If no movement occurs, you have a problem with or at the air vane; or in the mechanical type, an internal problem exists. An exception may be if the control arm is slipping on the shaft.

If the action is slow, the spring may not be reacting properly. Some installations have a number of holes in the control arm in which to fasten the spring. This changes the spring tension. If the spring is distorted or broken, always replace it with the correct type of spring. Proper tension is extremely important.

Some engines have an adjustment on the linkage to permit changing the length of movement and to compensate for wear. The incorrect spring tension may cause the engine to "hunt," that is, vary the speed. Try increasing the spring tension or adjusting the carburetor to eliminate the problem. Figure 9-3 shows a common linkage arrangement.

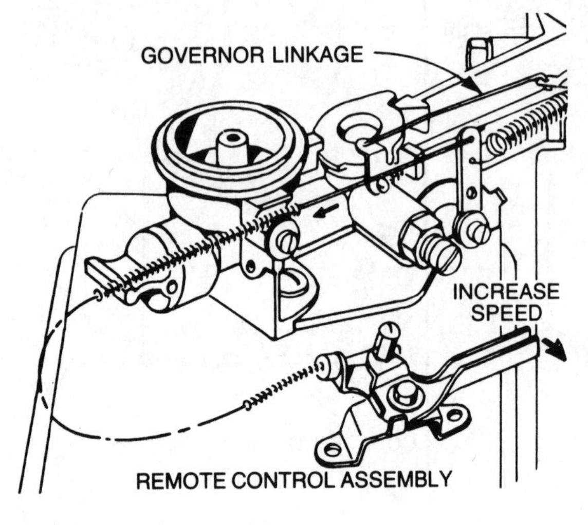

FIGURE 9-3
Governor linkage
Courtesy Briggs & Stratton Corp.

10 IGNITION SYSTEM

HOW THE IGNITION SYSTEM OPERATES
TYPES OF IGNITION SYSTEMS
TROUBLE SHOOTING IGNITION SYSTEM
SERVICING IGNITION SYSTEMS
SERVICING SPARK PLUGS
SERVICING BREAKER POINTS
IGNITION TIMING
CHECKING COIL AND CONDENSER

The internal combustion gasoline engine depends on an electrical spark to ignite the fuel charge in the combustion chamber. The spark must occur at exactly the right time in relation to the position of the piston. In addition the voltage must be high enough to make the spark jump across the spark plug gap against the resistance of compression within the combustion chamber. "High enough" means approximately 20,000 volts. To develop this spark and to deliver it to the combustion chamber, some type of ignition system is necessary.

A magneto ignition system is used on most small gasoline engines due to its simplicity and light weight. A battery ignition system, like the one found on an automobile, may be used on a few engines, but such a system requires a heavy battery and a means of recharging it. The magneto system makes use of a permanent magnet to produce electricity.

Despite variations in the design of the magneto ignition system, the basic operation is the same for all. Regardless of design differences all magnetos convert magnetic energy into electrical energy.

The electric current is generated by cutting magnetic lines of force, which are present in the permanent magnets, with a conductor. In this particular type of system, the conductor consists of a coil with many turns of wire. The rule-of-thumb is: The more lines of force (that is, the stronger the magnet), the more conductors (turns of wire), the faster the lines of force are cut (the speed) the more current that is produced. This whole sequence is brought about either by rotating a flywheel containing a magnet around a coil of wire or by rotating a coil of wire around a magnet. The results are the same in either case. Regardless of whether the ignition system is of the battery or magneto type, a low voltage is converted into a high voltage by self-induction.

Three general types of ignition systems may be found on small single-cylinder gasoline engines: (1) the flywheel magneto type, (2) the battery ignition type, and (3) the breakerless type, which can be either the battery or magneto type. Most engines are equipped with the flywheel type ignition system using breaker points. This system consists of (a) an ignition coil made up of a primary winding and a secondary winding mounted on a laminated core, (b) a condenser (capacitor), (c) breaker points (contact points), and (d) a spark plug. A cam or plunger driven by the camshaft or crankshaft is used to open the contact points. Most newer engines are equipped with the breakerless (solid state) type of ignition system.

HOW THE IGNITION SYSTEM OPERATES

An electric current is produced by rotating a permanent magnet, which is located in the flywheel, around the ignition coil. The magnetic lines of force surrounding the magnet are cut by the conductors—that is, by the wires that make up the primary coil—thus inducing a current of electricity in the coil wires. When electricity flows through a coil of wire, the coil becomes an electromagnet. The secondary coil is located within the magnetic field of the primary coil, and it is composed of many layers of very fine wire.

A set of breaker points, sometimes called "contact points," acts as a switch to open and close the primary circuit. The stationary point is grounded, and the insulated movable point is connected in the primary coil circuit. When the points are closed the primary circuit is complete, and the induced electricity, which was created by the magnet, builds a magnetic field around the secondary coil.

The breaker points are opened by a cam lobe or plunger driven by the crankshaft or camshaft. The points are closed by spring tension, which is part of the movable point. Since the two-stroke cycle engine fires every crankshaft revolution, the points are opened once every crankshaft revolution. In the four-stroke cycle engine, the cylinder is fired every other revolution, and so the points are usually opened by camshaft movement. When the points open, the primary circuit is broken and the magnetic field collapses through the secondary coil windings.

The condenser absorbs the surge of primary current, thus preventing the points from arcing. As the magnetic field collapses through the windings of the secondary coil, a high voltage is induced in the coil windings. At the same instant the charge stored in the condenser surges back into the primary winding, reversing the direction of current flow in the primary winding. This change of direction sets up a reversal in the direction of the magnetic field cutting through the secondary winding, which helps to increase the voltage in the secondary circuit. The pressure of this high voltage forces the charge to jump across the gap of the spark plug. This is the spark that ignites the air-fuel mixture.

TYPES OF IGNITION SYSTEMS

Although different arrangements of the components may be found on various engines, they all perform the same function in much the same manner.

The most common installation is the flywheel internal breaker, which has the coil, condenser, and breaker points mounted on a plate attached to the engine underneath the flywheel. The flywheel contains the magnets. Figure 10-1 shows a typical flywheel ignition system using an internal breaker point. This arrangement is compact and protected by the flywheel, but you must remove the flywheel to service the magneto. The breaker points are opened by a cam or plunger. On a two-stroke cycle engine, the breaker points can be opened directly by the crankshaft. The points on the four-stroke cycle engine are generally opened by the camshaft, which turns only half as fast as the crankshaft.

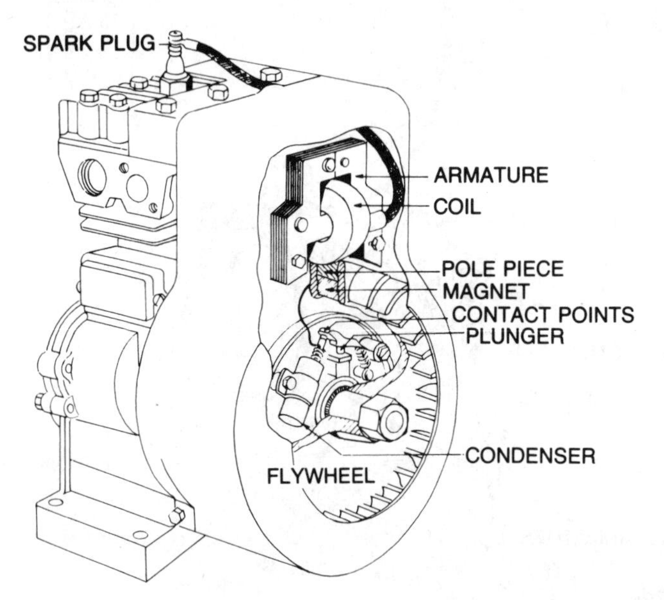

FIGURE 10-1
Flywheel ignition with internal breaker
Courtesy Briggs & Stratton Corp.

A variation of this arrangement has the coil mounted outside the flywheel. The breaker points and condenser are located on a plate covered by the flywheel (Figure 10-2). The breaker points may be operated by a cam or a plunger.

A third type of installation uses an external breaker system. With the coil assembly located above the flywheel and the magnet in the flyweel rim, the flywheel need not be removed to service the magneto unless the magnet itself is the problem. The breaker assembly and condenser are located on the outside of the engine near the camshaft (Figure 10-3). A plunger to open the breaker points extends into the engine where it contacts a lobe on the camshaft. The unit operates in exactly the same manner as the other two flywheel types with an internal set of breaker points.

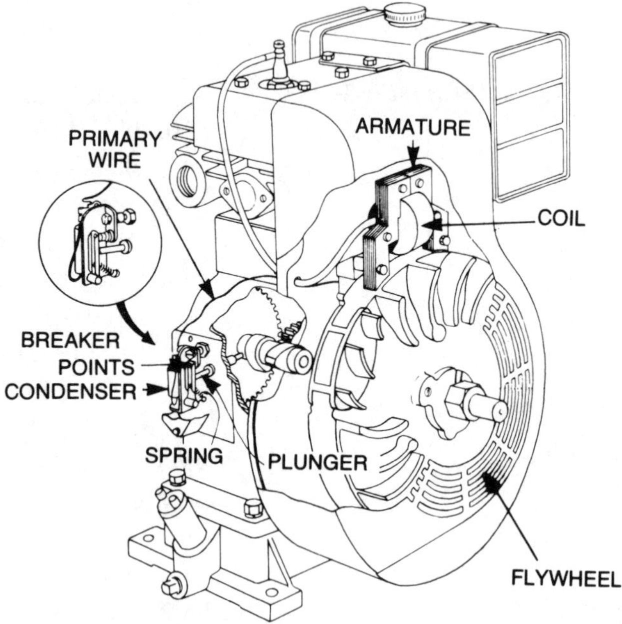

FIGURE 10-2
Ignition coil-mounted above flywheel
Courtesy Briggs & Stratton Corp.

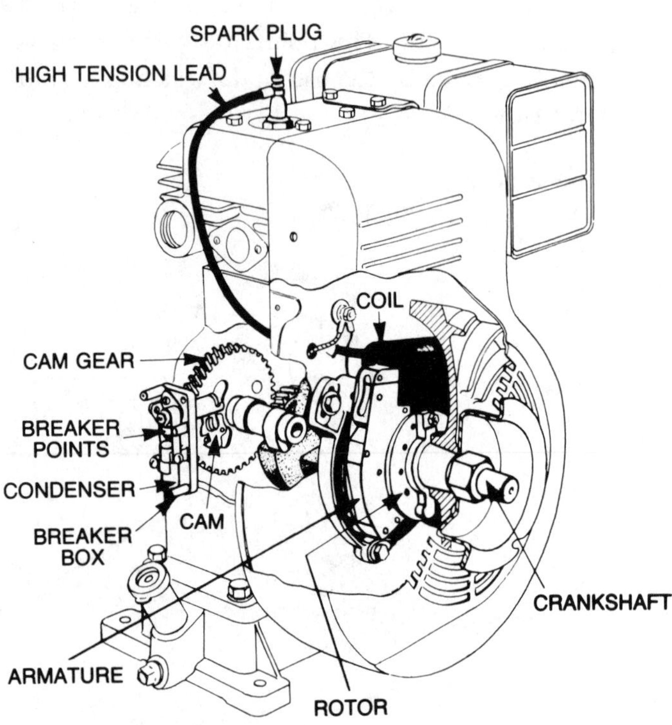

FIGURE 10-3
Ignition system with externally mounted breaker points
Courtesy Briggs & Stratton Corp.

Another variation with an external set of breaker points has a permanent magnet (rotor) attached directly to and turned by the crankshaft. The coil is mounted on an armature (core), which is attached to the side of the engine and covered by the flywheel (Figure 10-4). The armature helps to strengthen the magnetic lines of force.

A breakerless or solid state ignition system is used in some late model engines. The solid state system functions in the same general manner as the system using breaker points, but breaker points and a conventional condenser are not used. The only moving part is the flywheel. The solid state system includes four major components: (1) an ignition winding, (2) a trigger module, (3) an ignition coil assembly, and (4) a flywheel with a trigger projection. The system also includes the conventional spark plug and lead, plus the ignition switch (Figure 10-5).

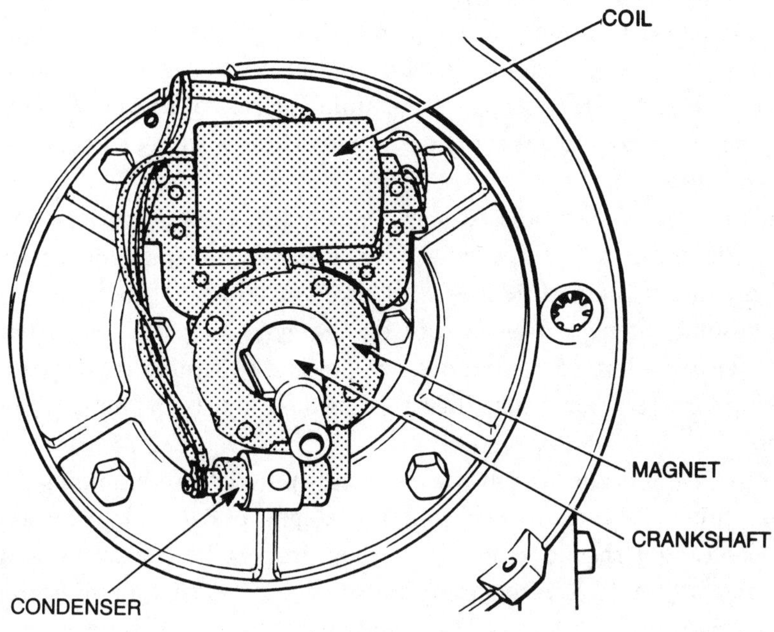

FIGURE 10-4
Ignition system with magnet attached to crankshaft
Courtesy Kohler Co.

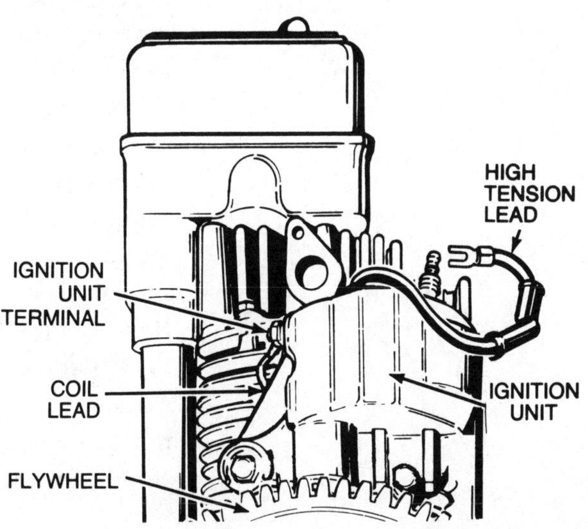

FIGURE 10-5
Solid state ignition system
Courtesy Tecumseh Products Co.

The trigger module, which serves the same function as the breaker points, includes diodes, a resistor, a sensing coil, and magnet plug which acts as an electronic switch. The ignition coil assembly includes a capacitor and a pulse transformer similar to a conventional coil. The flywheel has a special projection for triggering the spark to the spark plug. No service can be performed on this system other than testing and replacing parts.

A few engines may have a battery-operated ignition system, which functions in the same manner as the other types of ignition systems. The difference is that the source of energy for the coil is the battery rather than the revolving magnet.

TROUBLE SHOOTING IGNITION SYSTEMS

The only function of the ignition system is to deliver enough of a spark to the spark plug to ignite the compressed air-fuel mixture in the combustion chamber at the proper time. Three common problems that may be the result of malfunctions in the ignition system are: (1) failure to start, (2) misfiring, and (3) a loss of power. Always make the simplest and easiest checks first before working toward the more complex.

When the engine fails to start, check for a good spark to the spark plug. Such a test is very simple. Remove the spark plug wire from the spark plug. Hold the wire about 3/16 to 1/4 inch from a good ground, such as a clean spot on the cylinder head. Crank the engine with the switch on and observe the spark. If there is no spark, or if it is irregular or weak, then the problem exists in the ignition system.

Another way is to remove the spark plug before cranking the engine, thus enabling you to spin the engine easier and faster. In most cases, you should remove the plug, inspect its condition, and clean and adjust it if an engine problem exists. After loosening the plug a small amount but before removing entirely, clean away and blow out the dirt from around the plug. Then dirt will not fall into the engine when the plug is removed.

Another method to check for spark, as well as for the plug condition, is to set the spark plug gap at about 3/16 inch by bending the outside electrode. Replace the spark plug wire on the terminal. Hold the plug tightly on a clean spot on the cylinder so as to make a good ground. Then crank the engine with the switch on. A good spark should occur regularly at the plug gap if the ignition system is functioning correctly (Figure 10-6).

If there is no spark, or if it is irregular or weak, then problems exist in the ignition system. Check to make sure the switch is on and not grounding the system. Inspect the spark plug wire. Make sure that it is not cracked, or broken and that it has no indication of

FIGURE 10-6
Checking for spark and spark plug condition
Courtesy Clinton Engines Corporation

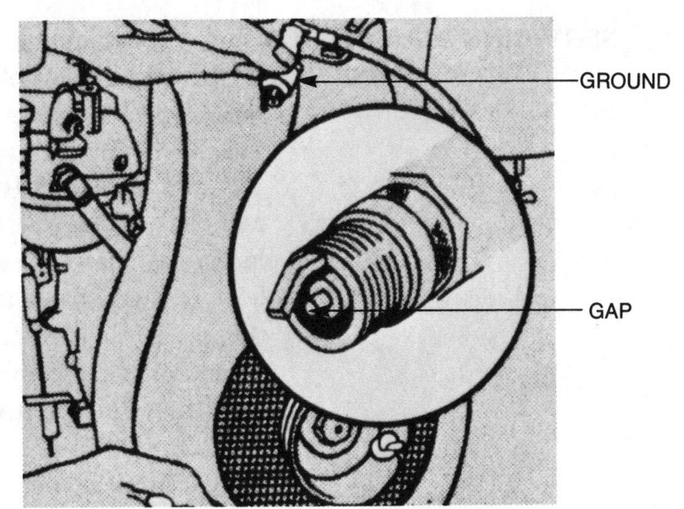

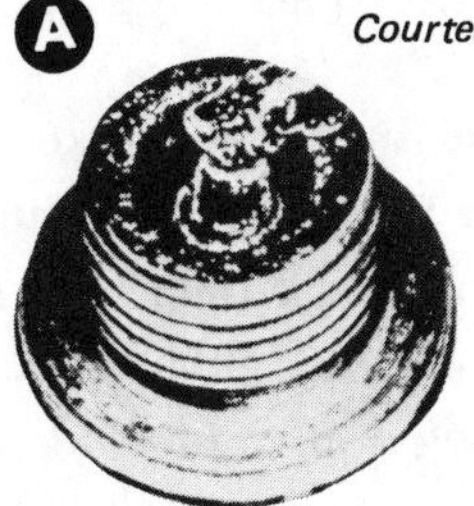

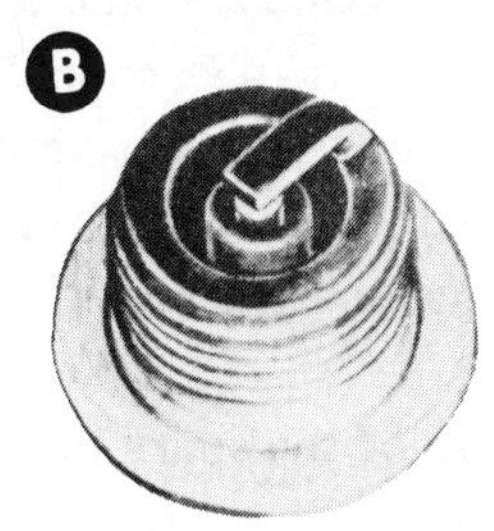

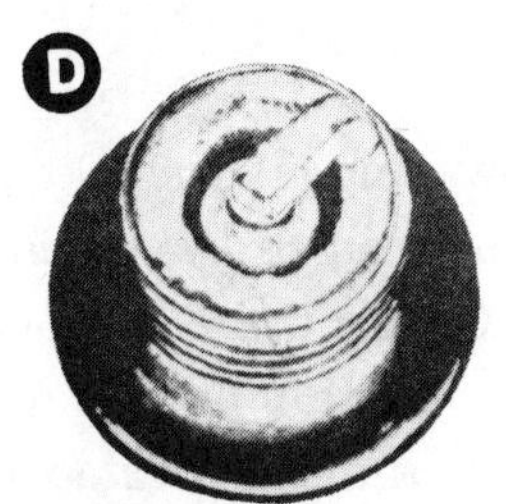

FIGURE 10-7
Spark plug conditions
Courtesy Harley-Davidson Motor Co., Inc. Subsidiary of AMF Inc.

arcing. Make sure the terminal fits tightly on the plug. If the wire and terminal are good, the problem exists in the ignition system and it needs to be serviced.

If a good spark is present, yet the engine fails to start, or if it starts but misses or lacks power, then check the spark plug. Carefully inspecting the spark plug can tell you many things about the condition of the engine, as well as about the spark plug itself. Spark plugs fail for various reasons. The insulator may be cracked or broken. The lower end of the plug may become coated with oil, carbon, or other deposits. This coating can result in the spark jumping from the center electrode to ground without jumping the spark gap. As the engine is used, the gap may erode so much that it is too wide for the spark to jump across, resulting in an engine miss.

Good operating conditions are indicated if the plug has a light brown to greyish tan deposit. A blistered white coating on the insulator indicates overheating. A black sooty carbon coating indicates an over-rich ruel mixture, which may be caused by a clogged air cleaner or an improper carburetor adjustment. An oily deposit can indicate that too much oil is getting into the combustion chamber. Eroded electrodes and a pitted insulator indicate a worn-out plug. If the spark plug is fouled or shorted, or if the gap is too wide, the engine may miss and be hard to start. A worn-out plug can result in the same problems. Figure 10-7 illustrates some types of spark plug conditions after extensive use.

SERVICING IGNITION SYSTEMS

The failure of the ignition system to deliver enough of a spark is an indication of trouble within the ignition. Except when the ignition breaker points and condenser are located on the side of the cylinder block, the flywheel must be removed to service the ignition system components.

SERVICING SPARK PLUGS

If the spark plug electrodes and the porcelain insulator are not burned away, cracked, broken, or blistered, the spark plug can be cleaned, its gap set, and put back into service.

If the plug is to be replaced, always use the same make and number or the equivalent. The correct heat range of the plug is important for good engine operation and long spark plug life. The heat range depends on the length of the porcelain insulator within the spark plug shell. The "reach" of the plug—how far the plug extends into the combustion chamber—is also important.

If the plug is to be reused, it should be cleaned. Using a pen knife, a wire brush, and solvent, remove all traces of carbon from around the porcelain and inside the metal shell. With an ignition point file, clean the end of the center electrode. This cleaning removes the scale and makes a clean flat surface where the spark jumps the gap.

If you have any doubt about the condition of the spark plug, replace it. Whether a new or old plug is used, the gap must be checked and set. Use a spark plug gauge, change the gap by bending the outer electrode (Figure 10-8).

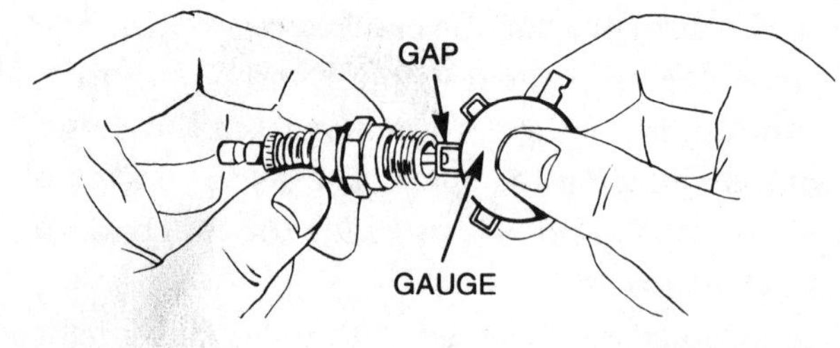

FIGURE 10-8
Checking spark plug gap
Courtesy Briggs & Stratton Corp.

A new gasket is supplied with a new plug. If the old plug is used, make sure the gasket is in good condition so the engine does not leak compression around the plug. Check the manufacturer's specifications for the gap setting. This specification can be found in the owner's manual or, in some cases, on a decal located on the engine. The recommended gap setting is usually 0.025 to 0.030 inch.

SERVICING BREAKER POINTS

If the breaker points are located on the outside of the engine, a metal cover is used to seal the unit from dirt and water. The cover is usually held in place with screws, which must be removed to service the points.

When the breaker points are located under the flywheel, the flywheel must be removed to service the system. On most engines you have to remove the blower housing for access to the flywheel. The rope-pull type of starting mechanism comes off with the housing, which is attached with bolts. A few engines may have an opening in the flywheel, which has a removable cover. Removing the cover permits you to inspect and adjust the points without removing the flywheel. Figure 10-9 shows the breaker points located inside the flywheel with a removable cover. This particular set-up is for a battery-type ignition system.

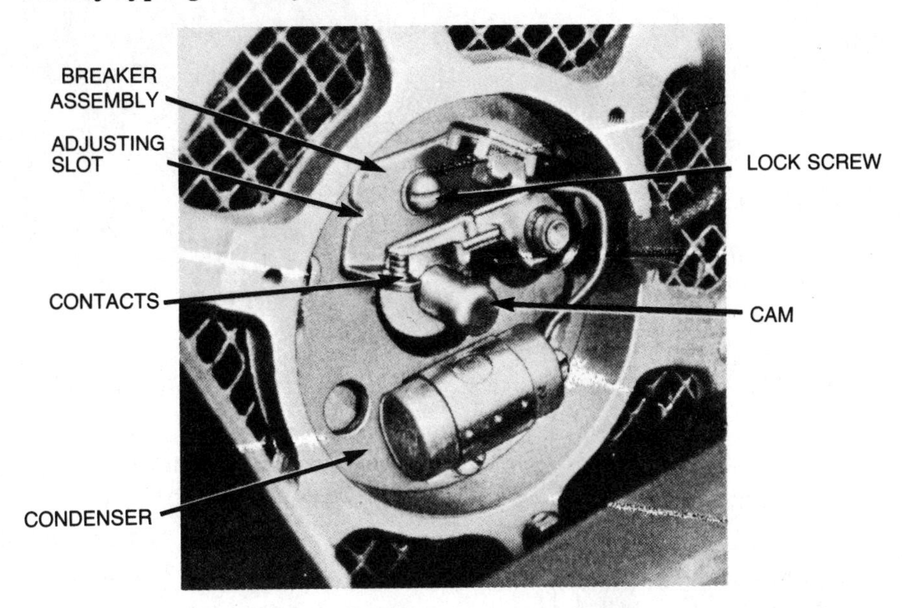

FIGURE 10-9
Circuit breaker and condenser used with battery system
Courtesy Harley-Davidson Motor Co., Inc. Subsidiary of AMF Inc.

The end of the crankshaft, on which the flywheel is mounted, is generally tapered using a key to keep the flywheel from turning on the shaft. A nut holds the flywheel in place. Because the shaft is tapered, once the flywheel comes loose from the shaft, it can be easily be lifted off. Most flywheels have provisions for attaching a puller to remove the flywheel.

If a puller is not available, you can, with care to prevent damage, remove the flywheel without one. Remove the nut that holds the flywheel to the crankshaft. Remove any washers that may be located between the flywheel and the attaching nut. Replace the nut, and screw it on until the top of it is exactly flush with the end of the shaft. An impact nut may be used in place of the regular attaching nut (Figure 10-10). Insert a heavy screwdriver or pry bar under the flywheel so you can pry upward. Be sure the screwdriver is prying on the flywheel and not on the plate under the flywheel. Hit the nut on the end of the crankshaft with a solid blow while prying upward. This blow should jar the flywheel loose. If it does not, rotate the flywheel a half-turn and repeat the operation. Take extreme care not to damage the end of the crankshaft. Always use a heavy hammer and hit the shaft and nut a solid blow. Do not just tap, or the threads may be damaged.

Check the breaker points. The points act as an automatic switch. On most installations an insulated movable contact point is connected to one end of the primary winding of the coil. A stationary

FIGURE 10-10
Loosening flywheel with impact nut
Courtesy Clinton Engines Corporation

FIGURE 10-11
Metal transfer (top and bottom) and contacts in good operating condition (middle)
Courtesy Kohler Co.

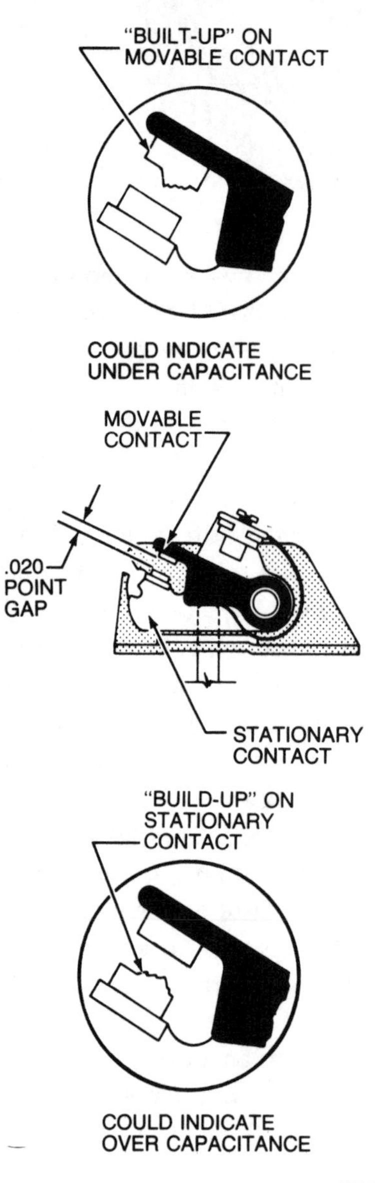

contact point is grounded. The points are opened by a cam or plunger and closed by spring tension. The distance that the point opens affects the amount of voltage that the coil can develop, as well as the ignition timing. If the point gap is too wide, the points open earlier and close later in terms of piston position. When the points open is when the spark is delivered to the spark plug. If the points are closed for too short a time, the coil produces a weak spark. If the breaker point gap is set too close, the points do not open until after the primary coil current has passed its maximum value.

In addition to gap setting, engine operation is greatly affected by the condition of the breaker points. If the points are burned or badly oxidized, little or no current can flow through. As a result the engine may not run at all or it may miss, particularly at full throttle. You can see that the adjustment of the point gap to specifications is very important and that the face of the points must not be pitted, oxidized, or burned blue. Properly functioning points have a grayish color; they are not pitted or oxidized. The point surfaces must contact one another squarely, face-to-face, to have a full flow of electricity. If you find it necessary to align the points, bend only the stationary point. If one point has developed a small pit and the other a projection, the contact surface can be smoothed by filing with an ignition point file. Remove only a small amount of material and keep the surfaces square. If the crater (pit) is deep, you loose part of the contact area when the projection is filed off. Install new points if the old points are badly pitted or if they have been operating misaligned so as not to have a full contact face. Figure 10-11 shows the metal transfer that can take place as a result of improper condenser capacity. A set of contacts in good operating condition is also shown.

Most breaker points have provisions for changing the gap setting. To check and set the gap, turn the crankshaft until the points are open to the widest gap. The gap is measured by inserting a flat feeler gauge using different thickness blades until the feeler blades just slip in between the contacts. Different means of adjustment are found on different installations, but the most common is an eccentric screw located in the breaker plate (Figure 10-9). Loosen the stationary point hold-down screw or screws. Turn the eccentric screw until you obtain the correct gap. Tighten the hold-down screws. On some installations, the gap is changed by slightly loosening the hold-down screws of the stationary point and prying the point one way or another until the correct gap is obtained. Retighten the hold-down screws.

Figure 10-12 shows one type of stationary contact adjustment using a hold-down screw. This type of point installation has one breaker point attached to the end of the condenser. The gap setting is changed by moving the condenser body back or forth after loosening the condenser clamp. Some breaker point gap settings are changed by turning an adjusting screw located in the stationary point bracket. A lock nut must be loosened before the adjusting screw can be turned. Other adjustment arrangements may be found, but they are usually self-evident.

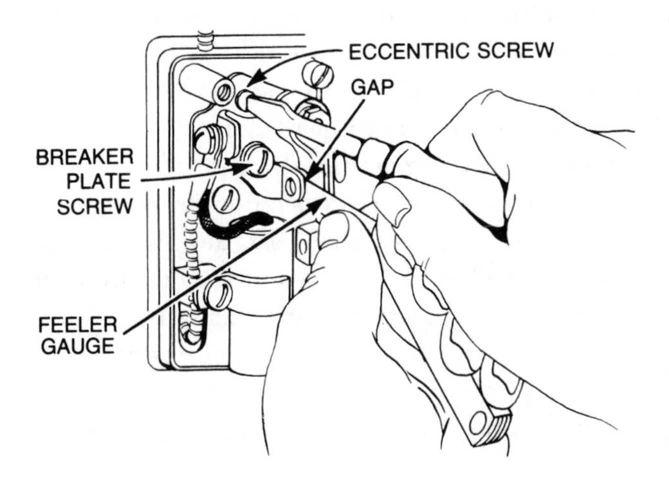

FIGURE 10-12
Breaker point gap adjustment
Courtesy Briggs & Stratton Corp.

Always check the manufacturer's specifications for the correct gap setting, most of which are approximately 0.020 inch. After changing the gap setting, close the points to make sure they are properly aligned. To obtain an accurate point setting the breaker point contact face must be smooth and true. The contact face must be clean and free from oil. Draw a piece of lintless paper between the points to make sure they are clean. Any oil or grease coming in contact with the points can cause trouble.

Be sure to check the fit of the plunger, if a plunger is used to open the points (Figure 10-12). Replace the plunger if it is loose in the bushing. Also replace the bushing if oil is leaking around the plunger.

IGNITION TIMING

Ignition timing refers to when the spark occurs at the spark plug. The spark flows to the spark plug the instant the breaker points open. This is when the fuel charge is ignited in the combustion chamber. Due to the momentum developed by the moving piston and crankshaft, the fuel charge is generally ignited a few degrees before piston top dead center (TDC).

Timing marks are generally found on the flywheel, and an indicator mark is usually located on some stationary part of the engine near the flywheel rim (Figure 10-13). The marks may indicate TDC and the point where the spark should occur. In some cases the only mark is where the spark should occur. The mark may be in the form of a line or a ball imbedded in the flywheel. Some engines do not have timing marks, but they require measuring piston travel in relation to TDC and when the points open. A few engines may use a timing pin, which is inserted through a timing hole. The engine turns until the pin slips into the hole in the flywheel. At this point the breaker points should begin to open.

In most cases carefully adjusting the breaker point gap results in correct timing. On some engines no timing adjustment can be made other than changing the gap setting of the points. The breaker point mounting plate can be shifted to change the timing on a number of engines. Normally, if the breaker plate has not been touched, the timing can be corrected by point adjustment alone.

Of various methods of checking ignition timing, the easiest and most accurate is by the use of an inductive neon or strobe light. In most cases you need a source of power when using a timing light for a magneto ignition system. Follow the manufacturer's directions for using such a light. Several types of neon or strobe lights are available, any of which can be used to check timing on a running engine.

This type of light is connected between the spark plug terminal and spark plug. The light flashes every time the plug fires. By aiming

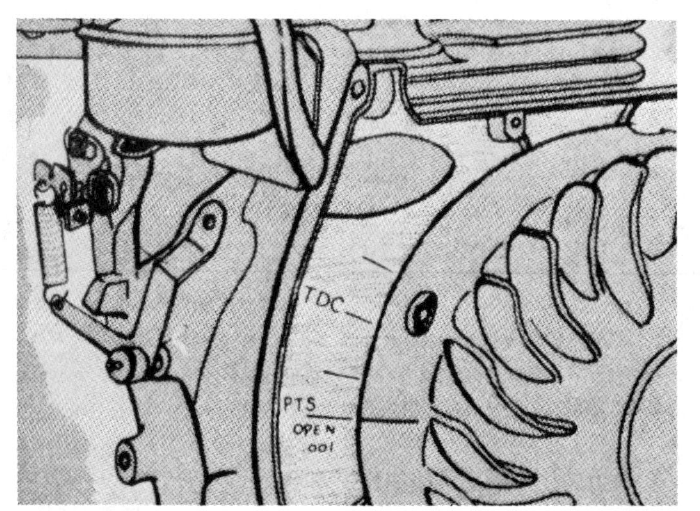

FIGURE 10-13
Timing marks
Courtesy Clinton Engines Corporation

the light at the timing indicator on the engine, you can see the mark on the flywheel when the plug fires, if the engine is properly timed. If it is not, make sure the point gap is correctly set and then shift the breaker plate or housing, whatever is adjustable, until the indicator and flywheel mark line up exactly even. Be sure to retighten all hold-down screws to lock everything in place.

If a timing light is not available you, can check timing by using an ohmmeter, a battery-operated test lamp, or a very thin piece of paper. The test is a matter of determining exactly when the points open in relation to the timing mark.

When using an ohmmeter or a test lamp, connect one lead to the movable breaker point and the other lead to ground. When the points are closed, the light should be on or the ohmmeter should show a specific reading for resistance. Disconnect the spark plug wire to prevent the engine from starting. Rotate the engine by hand in the direction of its normal rotation. As the points open, the light goes out, or the ohmmeter's reading changes. The timing mark on the flywheel should line up exactly with the indicator mark or with the center of the timing sight hole on some installations. Adjust the breaker points so they open at this point, if such is the method of timing, or move the breaker point plate to the housing to obtain the correct timing.

A thin piece of paper can be used in place of a light or ohmmeter. Place the paper between the breaker points. Exert a slight pressure on the paper as the flywheel is turned in the direction of its normal rotation. When the paper begins to pull out, the points are starting to open. The timing marks should be in alignment at this point.

Two-stroke cycle engines that obtain reverse by operating the engine in a reverse direction have two sets of timing marks, one for each direction. Point gap setting is critical for this arrangement.

A few engines may not have timing marks. TDC is obtained by checking piston position. A special gauge is used for this purpose after removing the spark plug.

CHECK COIL AND CONDENSER

Special electrical test equipment is necessary to test the coil as well as the condenser for ground, short, resistance, and capacity in the case of the condenser (capacitor).

If test equipment is not available, inspect the coil for evidence of overheating, cracked insulation, and broken or loose connections. The coil normally gives many years of satisfactory service. If all other possibilities have been eliminated, try installing a new coil.

In most cases replacing the coil involves checking and setting the clearance between the coil core (laminations) and the flywheel magnet. Before removing the coil check the clearance by inserting a flat feeler gauge between the coil core and the magnet. This test indicates the clearance you should adjust for when installing the new coil. As the air gap between the coil core and magnet becomes smaller, the magnetic lines of force become more effective. Under no circumstances should the magnet rub the core. Figure 10-14 shows how to check and set the magneto air gap when the coil is located outside the flywheel.

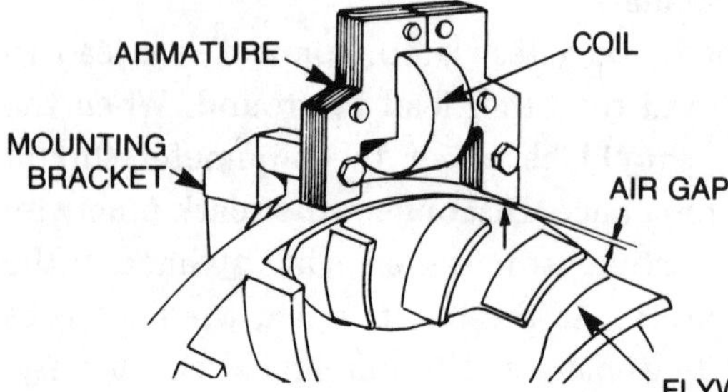

FIGURE 10-14
Checking and setting magneto air gap
Courtesy Briggs & Stratton Corp.

Check the strength of the magnet in the flywheel with a magnetometer. If you don't have such a device, you can make a rough test with a screwdriver. Place the flywheel upside down on a wood surface. Hold a screwdriver by the extreme end of the handle with the blade down. Move the screwdriver to within an inch of the magnet. The magnet should attract the screwdriver from this distance. If the magnet is too weak to do so, replace the flywheel.

The capacity and operation of the condenser are indicated by the condition of the breaker points. A faulty condenser causes the points to arc, become too hot, and turn blue. Excessive build-up (or transfer) of metal from one point to another indicates the wrong capacity condenser. It is generally considered good practice to replace the condenser when you replace the breaker points.

When the flywheel is removed, inspect all wiring to make sure that the connections are clean and tight, that the insulation is good, and that the wires are placed so they are not in contact with any moving parts. Make sure the key and keyway in the crankshaft are not worn or loose before installing the flywheel.

11 ELECTRICAL SYSTEM

PURPOSE OF THE ELECTRICAL SYSTEM
TYPES OF ELECTRICAL SYSTEMS
SERVICING THE ELECTRICAL SYSTEM
BATTERY MAINTENANCE
BATTERY TESTING
ELECTRIC STARTING MOTORS
CHECKING STARTING SYSTEMS
STARTING MOTOR DRIVES
CHARGING SYSTEMS
CHECKING CHARGING SYSTEMS
LIGHT SYSTEMS

There is a direct relationship between the ignition system and the electrical system. All small gasoline engines must have an ignition system to ignite the air-fuel mixture in the cylinder. The most common machines powered by small gasoline engines—such as walk-behind lawn mowers, rotary tillers, chain saws, and snow blowers—use magnetos to produce electricity for their ignition systems. They are started by hand, generally with a rope-recoil starting mechanism. Recreational machines—such as mopeds and the snowmobiles—have magneto-type ignitions. In addition, they may be equipped with lights and possibly tachometers. Electricity for these accessories is produced within the magneto assembly by using separate coils that develop a low voltage.

PURPOSE OF THE ELECTRICAL SYSTEM

The purpose of the electrical system is to supply electric power to operate the various components of the machine. The most common electrically operated unit is the starting motor, particularly on larger engines. Some lawn and garden tractors, as well as other mobile equipment, may also have lights, which in most cases are battery-operated. When they have electrical systems, a few engines use the system to supply electricity for the ignition system. A rechargeable battery, such as that used on an automobile, is the common source of power for the electrical system. Such a battery requires some means for recharging, which is generally accomplished by an engine-driven generator or alternator.

With a wide variety in the make-up of electrical systems, the following is only an overview of the major components that may be included in these systems.

TYPES OF ELECTRICAL SYSTEMS

The source of power for many electrical systems is a 12-volt battery, although some of the older engines use a 6-volt battery. The voltage of the system has little effect on the design or operation of the system, other than that all units must have a compatible voltage.

The most common device for starting the engine is a 12-volt starter/generator combination. The unit is mounted on the side of the engine and connected to the engine by a belt. When electric power from the battery is applied to the unit, it cranks the engine. The engine starts then drives the starter/generator, causing it to produce electricity to charge the battery.

A separate 12-volt starting motor, mounted on the side of the engine and using a retractable drive pinion to engage a gear on the crankshaft, may also be used to crank the engine. The same starting motor arrangement may be used on a few engines, only it requires a 110-volt power source. Such a motor is generally used on a stationary engine located close to a 110-volt power source. Some engines use a nickel-cadium battery power pack, which is recharged with a 110-volt rectifier.

A battery on a mobile machine needs some build-in means for recharging. As the battery is used, it loses its ability to provide electrical energy; unless recharged it reaches a point where it no longer functions. Recharging—which consists of sending a direct current through the battery—is generally accomplished by means of an alternator or a generator driven by the engine. A few installations may depend on a 110-volt battery charger, which is separate from the engine. The charger can be connected to the battery without removing the battery from the engine, but a source of 110-volt electricity must be available to operate the charger.

Of the several variations and combinations for charging batteries different engines, the combination starter-generator has widespread usage. The unit is wound internally so that, when driven by the engine, it produces a direct current to charge the battery. Some charging devices are called generators," while others are called "alternators." Both units perform the same function—charging the battery. They are also sometimes referred to as "DC generators" or "AC generators."

A separate generator or alternator may be mounted on the side of the engine and connected to the engine by a belt. In DC generators, a cut-out relay and a voltage regulator are incorporated into the circuit. The cut-out relay is to disconnect the generator from the battery when the engine is not running, and the voltage regulator is there to limit generator output. The alternator has either an internal or an external rectifying device in the charging circuit to change the alternating current to DC. A voltage limiter is also used in some alternator circuits to control the charging rate.

Some nonregulated alternators are in use, which have a built-in limited output such as 1½, 4, or 7 amperes. With the alternator thus tailored to the particular charging requirements of the system, you save the cost of a voltage regulator.

A dual circuit alternator, used on some engines, actually has two separate alternators located inside the flywheel. A magnet located in the flywheel provides the magnetic field for both sets of coils. One alternator has a rectifier in the circuit and provides DC for charging the battery; the other alternator supplies AC directly to the lighting system. The battery is not used for lights. Lights may be supplied only if the battery is disconnected and the engine is running.

In a magneto alternator, also used on some installations, the magneto supplies electricity in a regular manner for the ignition circuit. Additional and separate coils are used to supply electricity for lights.

When lights and other accessories are operated directly from the alternator or generator, the electricity must first go through a regulator. The regulator limits the current flow so the brightness of the lights remain reasonably steady rather than changing with the speed of the engine.

SERVICING THE ELECTRICAL SYSTEM

A few statements about the nature of electricity and its flow may help you to better understand the operation and the servicing of the electrical system.

To flow electricity needs a complete circuit (a circle or path) out from the source, through whatever unit is to be operated, and back to its source—be it magneto, alternator, or battery. Electricity travels out of one battery terminal through a circuit (wire) to the starting motor and back through the circuit to the other battery terminal. The same is true of the magneto: Electricity flows from the secondary coil terminal, out through the wire to the spark plug, where it jumps the spark plug gap and returns to the magneto.

Instead of using two wires, as in a house's electrical circuit, the metal parts of the engine and possibly the frame of the machine are used as a second wire. Electricity flows out from the source, into the circuit, and back to the source—all through the metal part of the engine and frame. So the circuit must be continuous through the wire to the unit to be operated (such as the starter motor or light) and then back to the source. Any break in the wire interrupts the circuit, and electricity cannot flow.

Because the metal parts of the engine and possibly of the machine itself are used as a second wire to complete the circuit, all the wires must be insulated from all the metal parts. The metal part of the circuit is known as "ground." One terminal of the battery is attached to the engine or to the machine itself, which is the ground side of the circuit. If the insulation on a wire carrying electricity is broken and touches the metal part of the machine, the circuit is said to be "short-circuited." The electricity returns to the source through the metal parts rather than through the starting motor or lights. Electricity, like water, follows the path of least resistance. It is easier for electricity to flow through the heavy metal frame than through a small wire.

A certain amount of electrical pressure (voltage) must be present at the source for electricity to flow. For that reason, the battery must be charged to force electrical energy through the wires. For the same reason, the magneto, alternator, or generator must be turning at a certain speed to produce enough voltage to provide that "pressure." Loose or dirty connections create a resistance to the flow of electricity in a circuit. Keeping these points in mind may help you to understand what must take place in a circuit for it to operate correctly.

The overhaul and repair of a starting motor, generator, alternator, or a combination starter/generator requires special test equipment. For the most part, very little service should be required other than replacing or adjusting the belt. If internal repairs are necessary, such as replacing the rotor, armature, or field coils, then replacing the entire unit is usually best. New or rebuilt units are generally available on an exchange basis.

No attempt will be made to go into the details of operating theory and the testing of the various units. Rather, information will be presented so you can become familiar with the various systems and identify the system used in your machine. In addition procedures will be developed for checking out the various units, permitting you to recognize and to isolate malfunctions that may occur.

Some basic test equipment is essential to determine specifically where the problem exists. Circuit continuity can be checked with a simple battery-powered test lamp. But for complete testing you must be able to determine the amount of current flow. So a low-range voltmeter for 0 to 16 volts, a low-range ammeter for 0 to 30 amperes, and an ohmmeter are essential for accurate electrical testing. A combination meter, which contains an ammeter, voltmeter, and ohmmeter, is available at a minimal cost at most automotive or electronics supply stores.

The battery, which is the heart of the electrical system, can be tested with a low-priced and unsophisticated hydrometer. By properly using a hydrometer you can determine the condition of the battery, as well as the state of its charge.

BATTERY MAINTENANCE

At the heart of the electrical system, the battery supplies, with few exceptions, the electricity for the starting motor and generally for the lighting system as well. Along with the charging system, the battery is also the source for the electrical system itself, as well as the source for battery-type ignition systems.

The commonly used battery is the lead-acid type, the same as used in automobiles. Most batteries used for today's engines are 12 volts, although some older engines use a 6-volt type. Certain characteristics are common to all lead-acid batteries:

1. The efficiency of the battery is lower in cold weather. As the temperature goes down, the battery efficiency drops.

2. A completely discharged battery freezes when the temperature goes below freezing. This ruins a battery.

3. A battery self-discharges if allowed to stand over a period of time without recharging.

4. The voltage of a battery is governed by the number of cells: a 12-volt battery has 6 cells; a 6-volt battery, 3 cells.

5. As a battery is charged, it produces hydrogen and oxygen gases, which explode if exposed to an open flame or spark.

6. Many of today's batteries are maintenance-free. Sealed so the gases cannot escape, these batteries do not require the adding of water. Batteries with vent holes in the removable cap allow the gases to escape. A battery of this type must be checked periodically, usually about every 100 hours of operation, to make sure that the top of the plates are covered with electrolyte. If the solution is low, barely covering the top of the plates, add water. Whenever possible add only distilled water; if not, use water with the lowest mineral content possible. Never overfill the battery, because as the solution expands when the battery is charged, the electrolyte overflows. Do not add water during freezing weather unless the engine is going to be used right away. The water freezes unless mixed with the electrolyte.

For long battery life the battery should be properly maintained:

1. Keep the electrolyte in each cell above the plates. There is generally an indicator (a protrusion) in the cell about an inch below the opening. Keep the solution near this level. If the battery needs water more than 3 or 4 times a year, the charging rate might be too high.

2. The top of the battery should be kept clean. An accumulation of dirt and moisture permits electricity to flow from the battery to ground. The top of the battery can be cleaned with a solution of baking soda and water—one tablespoon of soda to a quart of water. After cleaning, flush the top with water. Make sure the soda water does not get into the battery.

3. The battery carrier (rack) should be clean and the battery tightened down evenly and securely. Do not tighten to the point that the case cracks or distorts.

4. Be sure that the cables are in good condition and that the terminals and battery posts are clean and tight. The battery posts and the insides of the terminals can be cleaned with sandpaper. After cleaning, grease the posts and terminals lightly with petroleum jelly to retard corrosion. Make sure the ground cable is clean and tight at the engine block or frame.

5. Check the polarity of the battery to make sure the battery is not reversed. Most systems use negative (−) for ground.

6. Make sure the vent holes in the caps are kept open.

7. When removing the battery terminal from the post, do not use force since you may break the post loose from the plates. After loosening the bolt or clamp, spread the terminal with a pliers or screwdriver and lift it off. Never hammer on the posts.

8. A battery should not stand idle for a long period of time, such as 6 or 8 weeks, without being charged for at least a short time. When a battery is not used for a period of time, the plates sulfate, which reduces the capacity of the battery. A trickle charger with an output of about 2 amperes may be used to charge the battery for at least 12 hours, or the engine can be started and run for a half-hour or more. The nickel-cadium battery pack, having its own 110-volt charger, is rechargeable. For best service the battery should be charged for a day before the engine is to be used. The battery can be left on continuous charge between operations.

9. When replacing the battery, always select a replacement with the same or better electrical specifications in terms of ampere-hour capacity, which is important for cold weather starting. If in doubt, a battery with a 50-ampere-hour capacity serves almost any load requirement.

BATTERY TESTING

The state of the charges, as well as the condition of the battery, can be checked with a battery hydrometer for a battery having removable cell caps. The state of charge cannot be checked on a maintenance-free battery having a sealed cover. But these batteries have a charge indicator located under the top of the cover beneath a sight glass. If it is necessary to check its condition, make a high-rate discharge test on this battery.

A hydrometer is available at a reasonable cost from almost any automotive parts and accessory store. The level of the electrolyte must be above the plates to use a hydrometer. If you have to add water to bring the electrolyte to its proper level, the battery must be charged for several minutes to thoroughly mix the electrolyte.

To test the battery remove the cell caps and draw electrolyte from one cell up into the hydrometer, just enough so the float rides free. Hold the hydrometer at eye level, and read the scale. The reading is taken on the float at the surface of the liquid (Figure 11-1).

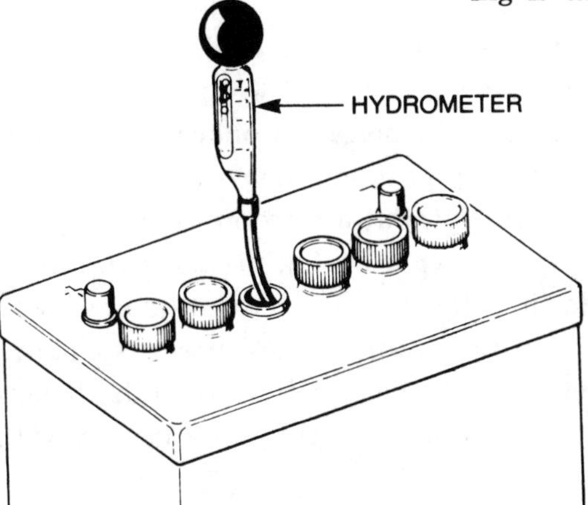

FIGURE 11-1
Testing the battery with a hydrometer
Courtesy Tecumseh Products Co.

Return the liquid to the cell. Test the rest of the cells in the same manner. Compare the readings of each cell. If one or more cells show a specific gravity deviation of over 0.025 points, the low cells are faulty and the battery should be replaced. If the cells are all below 1.225 (half-charge), charge the battery and retest it. A fully charged battery is from 1.260 to 1.280. A 25% charged battery is approximately 1.170 to 1.200. A battery with a specific gravity reading below 1.150 is considered to have no useful purpose. When the specific gravity reading is low, charge the battery with a slow charger and retest it. If the battery is still 1.225 or less, the battery is not taking a charge and should be replaced. The quick battery charger found at most service stations can be set to give a battery a slow charge over a period of 12 or more hours. If you have a doubt about the battery and if it comes up to 1.250 or better after charging, let it sit overnight and then retest it to see if it holds a charge.

If the battery is maintenance-free and does not have a charge indicator, it must be checked with a voltmeter. Numerous voltage testers are on the market for testing the voltage of the complete battery while under load. A no-load voltage test is of little value other than showing a completely dead cell. With a hard cover on the battery, the individual cell voltage cannot be measured with an ordinary voltmeter.

The high-rate discharge test, the most common, indicates the ability of the battery to operate under load. All garages and most service stations have this type of tester. It generally includes an ammeter to show the load imposed, a variable resistor to place a load on the battery, and a voltmeter.

Before attempting to make this test, fully charge the battery. The leads are attached to the appropriate battery posts, positive to positive and negative to negative. A load of 3 times the ampere-hour rating is applied by the variable resistance: approximately 150 amperes, or 50 ampere-hour battery times 3. The voltage of a serviceable battery should be at least 9.6 volts. If not, replace the battery. Do not apply the load for over 15 seconds. In the case of the 6-volt battery, the voltage should be at least 4.8 volts under load.

If you have only a voltmeter with a 0- to 16-volt scale, you can still get an idea of the condition of the battery. Connect the positive voltmeter lead to the positive battery post and the negative voltmeter lead to the negative battery post. Without a load on the battery the voltmeter should read at least 12.6 volts. Using the same voltmeter hook-up, turn the headlights on, if the machine is thus equipped, and crank the engine with the starting motor. If the battery is charged and serviceable, the voltmeter should read at least 10 volts. If not, replace the battery. Figure 11-2 shows the testing of a battery with a voltmeter.

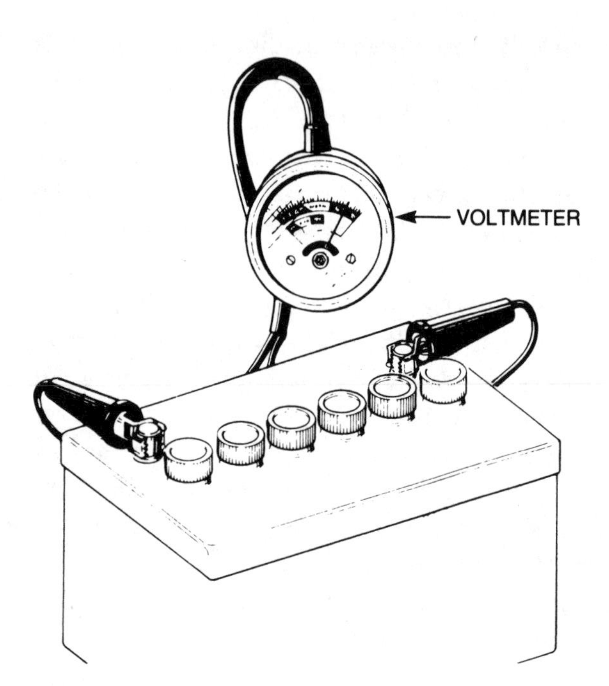

FIGURE 11-2
Testing the battery with a voltmeter
Courtesy Tecumseh Products Co.

If in doubt about the capacity of the battery, have it fully charged and have a high-discharge test made. This procedure definitely determines the condition of the battery.

ELECTRIC STARTING MOTORS

Of the different starting motors on engines, most are used on the larger type lawn and garden tractors. A number of engines are equipped with a combination starter/generator. Also, the single-unit starter operates on a 12-volt battery or, on older engines, on a 6-volt system. In some cases an 110-volt AC-DC starter is used. This starter is very similar to the 12-volt DC individual starter, only the source of electricty is supplied by a 110-volt AC power source, which is rectified to 12 volts DC.

Let's begin with a brief general description of the direct current starting motor. A starting motor should not be operated continuously for over 10 seconds. Allow a full minute between each 10-second cranking interval, and do not crank the motor more than 5 times. Instead, check for engine malfunction. This rule applies to all starting motors including the combination starter-generator.

The starting motor consists of a steel cylindrical housing, to whose inside steel pole shoes are attached. Generally two field coils are made up of a number of turns of wire, wound to fit around the pole shoes but insulated from both the starter housing and shoes. The coils are connected in series with the armature by brushes and directly to the starter terminal. Some starting motors use only permanent magnets in the housing in place of wound field coils. The brushes act as contacts to transmit electricity from the battery to the commutator on the armature, as well as to the field coils if wound field coils are used. The brushes are held in contact with the commutator by spring tension. Figure 11-3 shows a disassembled starting motor.

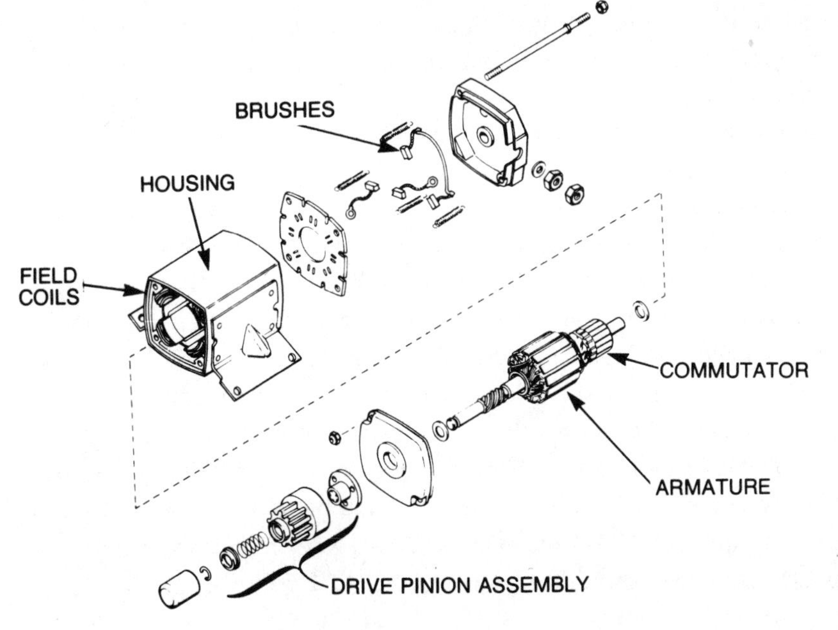

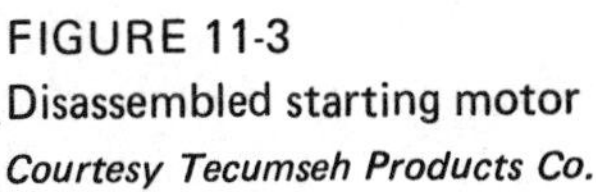
FIGURE 11-3
Disassembled starting motor
Courtesy Tecumseh Products Co.

The armature consists of a slotted iron core, a shaft for supporting the unit, and a commutator. The slots hold coils of insulated wire, the ends of each coil being soldered to insulated bars on the commutator. The commutator is insulated from the shaft and core. The brushes, which ride on the commutator, consist of two ground and two connected to the field coils; the latter two transfer electricity to the commutator.

An end cap (housing) supports the commutator end of the armature shaft on a bushing. The drive housing, in which the gear drive assembly is supported, supports the other end of the starter shaft. When a pully is used for a drive belt, the pulley is mounted on the shaft that is supported in the housing by bushings.

In general, the starting motor operates on the principle of magnetic repulsion. A magnetic field is created in the armature windings when the battery is connected to the starting motor. Another magnetic field is created in the field coils at the same time. The commutator is used to change the direction of current flow as it revolves. This change of direction reverses the magnetic polarity. Because like magnetic poles repel one another, as the direction of current flow is changes, so is the magnetic polarity. This ongoing reversal causes the armature to rotate.

CHECKING STARTING SYSTEMS

A number of malfunctions may occur in the starting system: (1) The starting motor may not crank the engine. (2) The starting motor may turn but not crank the engine. (3) The starting motor may crank the engine vary slowly.

If the starting motor does not crank the engine, first check the condition of the battery. The battery should have a specific gravity reading of at least 1.225 (half charge). Chances are that, at this charge, the starter motor is not able to crank the engine at as high a speed as with a fully charged battery. If the battery is low on charge, either recharge or replace it before making further tests. Figure 11-4 shows a complete battery-starter circuit for a motor generator unit. The battery is used for ignition on this installation, and the regulator controls the generator charging rate.

Check the terminals at the battery posts. They must be clean and tight. If you have any question about whether they are clean, removing them from the battery posts and cleaning them with sandpaper is best. If you have a 0- to 5-volt range voltmeter calibrated in tenths, place one lead on the battery post and the other on the terminal. Try to crank the engine with the starting motor. If the voltmeter reads 0.02 or more, there is a resistance in the connection. The terminal must be removed from the post and cleaned. This procedure is known as a "voltage drop test" and can be used to check any electrical connection if electricity is flowing. Connections other than battery terminals should not have over a 0.01-volt drop.

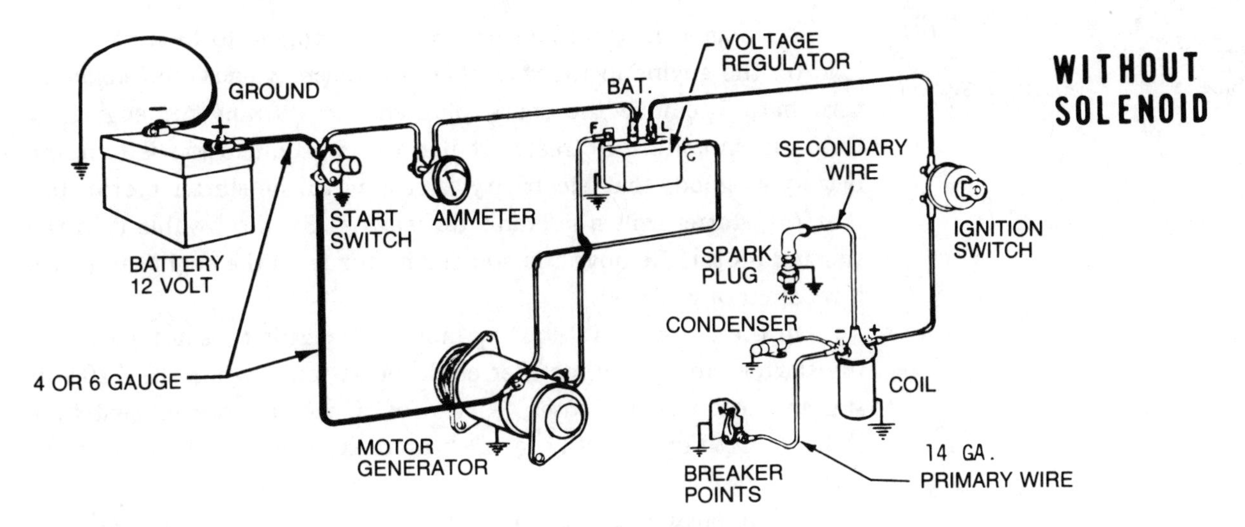

FIGURE 11-4
Motor-generator circuit
Courtesy Kohler Co.

Check the ground cable connection on the frame or engine, which must be clean and tight. Check the cables to the switch and starter. If the connections are loose or dirty, then remove, clean, and retighten them. Feeling the connections with your fingers may help to detect a poor connection. A dirty and/or loose connection may feel warm to the touch when you are either cranking the engine or trying to crank it.

If the battery is good and the cables are in good condition—clean and tight—check the starter switch. To do so, connect a cable (a heavy wire) directly from the battery to the starter motor, or bypass the switch by connecting the two starter switch terminals together with a heavy conductor. If the starter now operates, the switch is at fault and should be replaced.

A voltmeter with a 0- to 15-volt range can be used to check voltage to the starter motor. Place the negative voltmeter lead on the starter motor housing, and place the positive lead on the starting motor terminal. Operate the starter switch. The voltmeter should read the battery voltage. If you lack a voltmeter, disconnect the starter cable at the starter. While holding the starter switch closed, scratch the starter cable across a good ground. A very heavy and bright spark should etch the metal. (Keep the arc away from the gasoline!) If there is battery voltage at the starting motor housing or a good spark to the housing, place the voltmeter positive lead on the starter terminal and close the starter circuit. If the internal circuit in the starting motor is complete, the voltmeter should read battery voltage. Lacking a voltmeter, just touch the starter cable to the starter terminal for an instant. If there is an arcing the starting motor circuit is complete. If there is no voltage reading or arcing, the internal starting motor circuit is open and the starter must be removed from the engine.

Another possibility is that the starter drive pinion may wedge into the gear on the engine, so the starter cannot turn. Loosening the starter from the engine should allow the gear to slip back into place.

An engine malfunction can cause the engine to lock. And any load on the engine in the way of equipment may cause the engine to turn hard. Remove the spark plug and try turning the engine by hand. It should turn freely. If it does, the conclusion is that the battery is good, that electricity is getting to the starter motor, but that the starter will not crank the engine. So the trouble is in the starting motor. Remove the starting motor from the engine and have it repaired or replaced.

In the 110-volt AC starting motor, if electricity is not getting to the starter motor, the rectifier could be at fault. When the 110-volt starter motor blows fuses, check for a shorted starter switch, a shorted rectifier assembly, or a shorted extension cord to the rectifier.

If the engine cranks slowly, the battery may be discharged, perhaps due to faulty connections or cables. The lubricating oil in the engine may be too heavy if the weather is cold. The load may not be completely released from the engine. Remove the spark plug and turn the engine by hand to see whether the engine can be turned freely.

A dirty or worn commutator, worn brushes, or weak brush holder springs may cause the engine to turn over slowly. If so, the starting motor must be removed from the engine to service. If the armature is rubbing the field coils, or if it is binding due to worn bushings, binding bushings, or a bent shaft, then the unit should be removed and either repaired or replaced.

If the starting motor spins but does not crank the engine, the drive pinion gear may be sticking. Washing the pinion gear and shaft in a petroleum-type cleaner generally frees up the pinion so it spins in and out freely on the shaft. The belt may be slipping if a belt drive is used. The ring gear on the engine and/or the drive pinion on the starter drive may be damaged.

A free-running test gives an indication of the internal condition of the starting motor. A 12-volt battery and a 0- to 25-ampere ammeter is necessary. Place the starter motor in a vise to hold it steady. Attach one end of a heavy cable, like an automotive booster cable to the negative post on the battery, and attach the other end to the starter housing. Attach the end of other cable to the positive battery post and the opposite end to the ammeter. Attach a heavy cable to the other connection on the ammeter, and attach this cable to the terminal on the starting motor. If the starter motor is in good operating condition, it should turn at a high rate of speed and not draw more than 6 amperes.

STARTING MOTOR DRIVES

The single-unit starting motor uses a drive assembly to turn a pinion gear, which engages a ring gear. The ring gear is attached to the engine crankshaft or flywheel. The drive assembly is constructed so as to prevent the starting motor from being driven by the engine when the engine starts. The starting motor armature shaft has splines upon which is mounted a helix gear; in some cases, a helix-type thread is machined on the shaft (Figure 11-3).

When the starting motor is activated, the pinion gear should turn on the threads of the drive assembly engaging the ring gear and cranking the engine. As soon as the engine starts, the ring gear revolves faster than the starting motor, thus driving the pinion on the armature shaft out of engagement with the starter ring gear.

The starter drive pinion must spin freely in and out on the armature shaft to function properly. If not, clean the assembly in a petroleum-base cleaner. Do not wipe it entirely dry, because the cleaner acts as a lubricant. If the pinion gear still does not move freely back and forth on the shaft, or if the gear teeth are worn or chipped, replace the entire assembly. The assembly may be held on the starting motor armature shaft by a nut or retaining clip.

CHARGING SYSTEMS

The purpose of the charging system is to recharge the battery. The system may be made up of a direct current (DC) generator, a combination motor generator, or alternator, which is an alternating current (AC) generator.

The charging system using a DC generator is made up of the battery, a belt-driven DC generator, and a regulator assembly containing a voltage regulator, a cut-out relay, and possibly a current regulator.

When a motor (starter)-generator combination is used, the windings in the motor-generator are heavier than those used in a single-unit generator. The circuit in the motor-generator is different from that used in the unit generator. The motor-generator unit is contained in one housing. A starter switch is added to the circuit. A solenoid may or may not be used; when it is used, the starting motor can be actuated with the ignition key or a pushbutton switch. Without a solenoid, the starter switch has heavy contacts and must be manually actuated, generally with the foot. Figure 11-5 shows a motor-generator circuit with a solenoid.

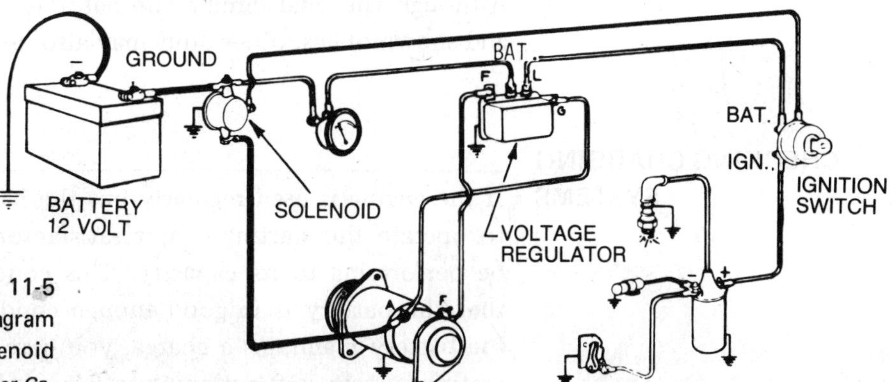

FIGURE 11-5
Motor-generator wiring diagram with a solenoid
Courtesy Kohler Co.

Like the single-unit DC generator, the alternator system can be a separate unit. But the most common alternator systems use a permanent magnet attached to the inner rim of the flywheel, like the magneto. A stator assembly is attached to a plate on the engine, which is located behind the flywheel. The magneto usually has only one stator coil as the source of energy. When an alternator is used to charge the battery, the current produced is alternating (AC) in nature. A rectifier or diode arrangement of some sort must be used in the circuit to convert the AC electricity to direct current (DC) before going to the battery, which can be charged only with direct current. Figure 11-6 shows the coil arrangement used to provide electricity for the ignition system, along with an alternator to charge the battery or to operate a lighting system.

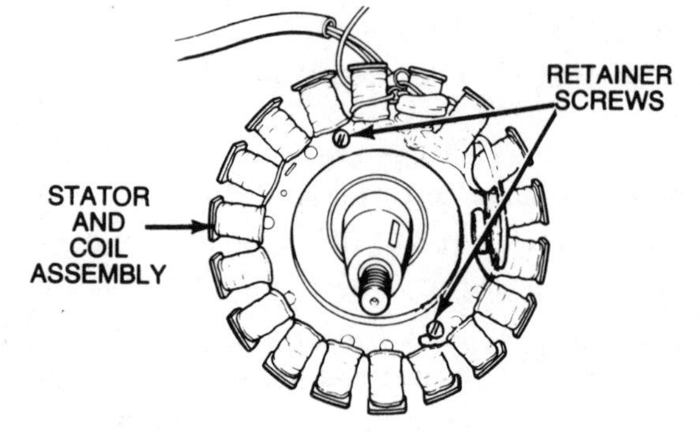

FIGURE 11-6
Combination ignition coil and alternator coils
Courtesy Tecumseh Products Co.

A voltage regulator is used in many alternator systems to limit the charging rate according to battery requirements. Figure 11-5 shows the regulator in the circuit.

A dual circuit alternator, used on some engines, is basically two alternators in one. One alternator system uses a rectifier in the circuit to supply electricity to charge the battery. The output is limited through the construction of the coils in the stator; no regulator is needed. The alternator system supplies alternating current for the operation of the electrical accessories, typically headlights. Current is available as long as the engine is running. The brightness of the lights varies with the engine speed. The outputs from the two units are separate, so the load imposed on one does not affect the other. Although the dual circuit alternator is commonly used on mopeds and snowmobiles, other units may also use this arrangement.

CHECKING CHARGING SYSTEMS

If the engine is used regularly but the battery is not charged enough to operate the starting motor satisfactorily, the generator may not be performing to its capacity. This conclusion, of course, assumes that the battery is in good enough condition to accept a charge. If the battery maintains a charge, you may safely assume that the generating system is functioning satisfactorily. On the engine equipped with a dual circuit alternator, if the battery is kept charged and the lights operate satisfactorily, the charging system does not require service.

Some engines are equipped with an ammeter, which indicates whether the generator or alternator is charging—and *how much.* (Generators may charge but not at a high enough rate to take care of the electrical load.) If the charging system is equipped with a regulator, you adjust the charging rate to accommodate an increased load. When the generator output is too high, the battery may require an excessive amount of water.

With a minimum of tools and equipment, you can locate the charging problem if you understand what is expected of the system. Although commercial establishments use a generator-regulator tester, which consists of an ammeter, voltmeter, a fixed resistor, and a variable resistance, you can locate problems with a lot less equipment.

You can make a test lamp from a 12-volt bulb holder and bulb connected to two leads with alligator clips. Use this lamp to check current flow in a circuit, as well as the circuit itself.

An ammeter with a 0- to 10-ampere range is helpful. A multimeter containing a voltmeter, ammeter, and ohmmeter can be obtained from a radio and electronics parts supplier at a reasonable cost. This meter enables you to make many electrical tests without a separate ammeter or test lamp. Figure 11-7 illustrates a charging circuit using a voltage regulator.

Before making any charging system tests, make sure that the battery is fully charged and that the terminals are clean and tight. Check the complete wiring circuit to make certain that there are

FIGURE 11-7
Voltage regulator and charging circuit
Courtesy Harley-Davidson Motor Co., Inc. Subsidiary of AMF Inc.

OPERATION:

CENTER MOVABLE CONTACT WORKS BETWEEN CENTER POSITION AND UPPER CONTACT FOR HIGH CHARGE RATE OR BETWEEN CENTER POSITION AND LOWER CONTACT FOR LOW CHARGE RATE.

1. LOW VOLTAGE—UPPER CONTACTS CLOSED—FULL FIELD CURRENT
2. MEDIUM VOLTAGE—BOTH CONTACTS OPEN—PARTIAL FIELD CURRENT THRU RESISTOR
3. HIGH VOLTAGE—LOWER CONTACTS CLOSED—NO FIELD CURRENT (SHUNTED FIELD)

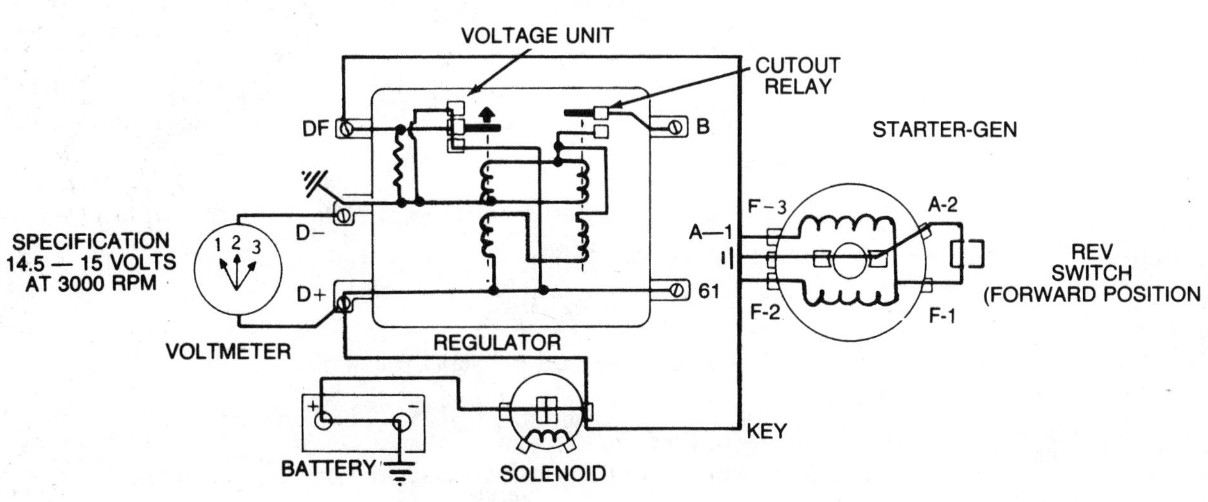

no shorts or broken wires and that all connections are clean and tight.

If the generator or alternator is belt-driven, check the belt tension. The belt should be adjusted so there is 1/4 inch movement up and down when the belt is pressed with thumb pressure midway between the pulleys. The belt is adjusted by moving the generator in or out in the adjusting slot after loosening the adjusting bolts (Figure 11-8).

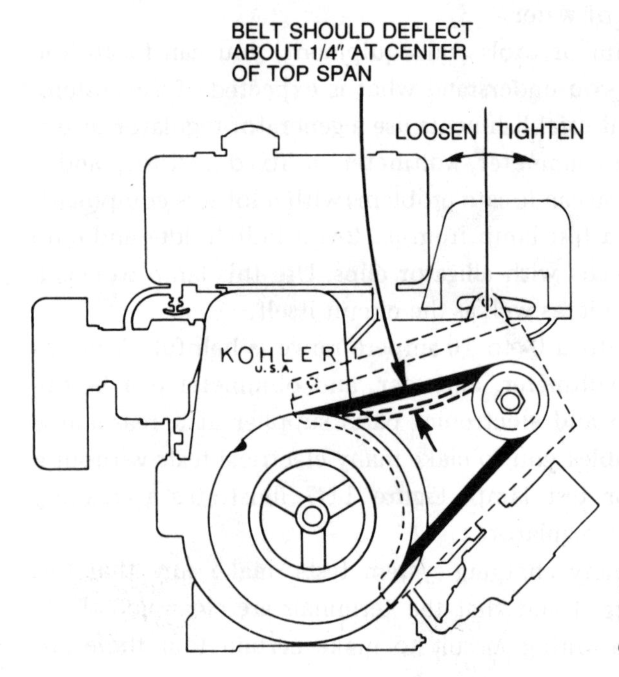

FIGURE 11-8
Belt tension check and adjustment
Courtesy Kohler Co.

Checking the charging system is a matter of isolating the problem with a minimum amount of disassembly. Examine the basic possibilities first:

1. Check the battery first to determine if it has a charge and if it will hold a charge.

2. Check the battery terminals to make sure they are clean and tight. Make sure the ground cable is clean and tight where it attaches to the engine or frame, because a resistance in the circuit affects the charging system.

3. Check the insulation on the wires, especially for bare spots where shorts may occur.

4. Check for broken wires.

5. Check for loose and/or dirty connections.

6. Make sure all components have a satisfactory ground.

7. Check the switch carefully to make sure it is functioning.

To determine if the generator or alternator is producing electricity, disconnect the wire from the regulator that goes to the battery. If a regulator is not used, disconnect the wire between the generator and the battery. Insert a test lamp into the circuit. If an ammeter is available, insert it into the circuit in place of the test lamp. Start the engine and run it at a normal operating speed. The test lamp should light if electricity is being produced by the generator or alternator. As soon as the bulb lights, disconnect the test lamp and shut off the engine. The bulb may burn out if left in the circuit. If an ammeter is used, the meter indicates the rate of charge.

A simple rule in using an ammeter is that the circuit must be broken and the ammeter inserted into the circuit. In other words, an ammeter is said to be connected "in series" in a circuit. A voltmeter is connected "parallel" in a circuit, that is, one of its connections is grounded.

If a regulator is used, disconnect the wire from the "F" terminal of the regulator. If the generator output remains the same, the generator is defective. Then ground the generator "F" wire. The generator charging rate should increase. With a grounded "F" field connection, the generator should charge at its maximum. If there is no change in the charging rate when the "F" field is grounded, the generator is at fault.

If the unit is a combination starter/generator system, make sure the cable between the battery and starter switch is clean and tight. Do the same for the cable and connection between the switch and starting motor.

If the generator or alternator is not functioning, it should be removed from the engine. The overhaul, testing, and rebuilding of the generator and/or alternator, as well as the adjustment of the regulator unit, require special equipment. Many service establishments do not service the units but simply exchange them for rebuilt ones.

LIGHT SYSTEMS

A lighting system, used on some machines, may include just one or two headlights, or it may include headlights, instrument lights, tail lights, and stop lights. Two different systems may be used:

1. One system is operated by the alternator, and no battery is required. Figure 11-9 is a diagram of such a lighting and ignition system. Since the engine must be operating to have lights, the lights vary to some extent according to the speed of the engine. A regulator is incorporated in most of these installations to limit the amount of voltage reaching the lights. Too high a voltage causes the bulbs to burn out. This type of lighting system is generally found on mopeds and snowmobiles.

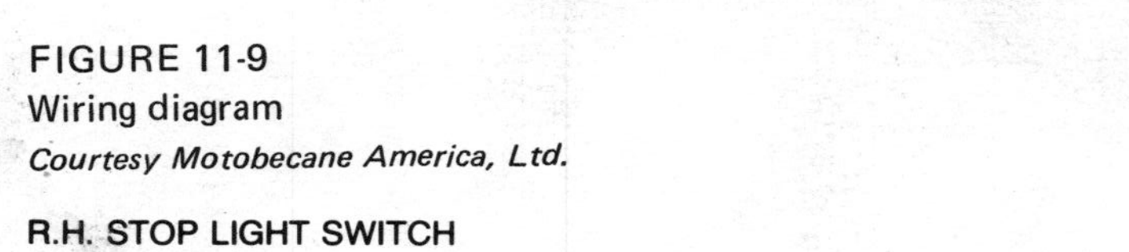

FIGURE 11-9
Wiring diagram
Courtesy Motobecane America, Ltd.

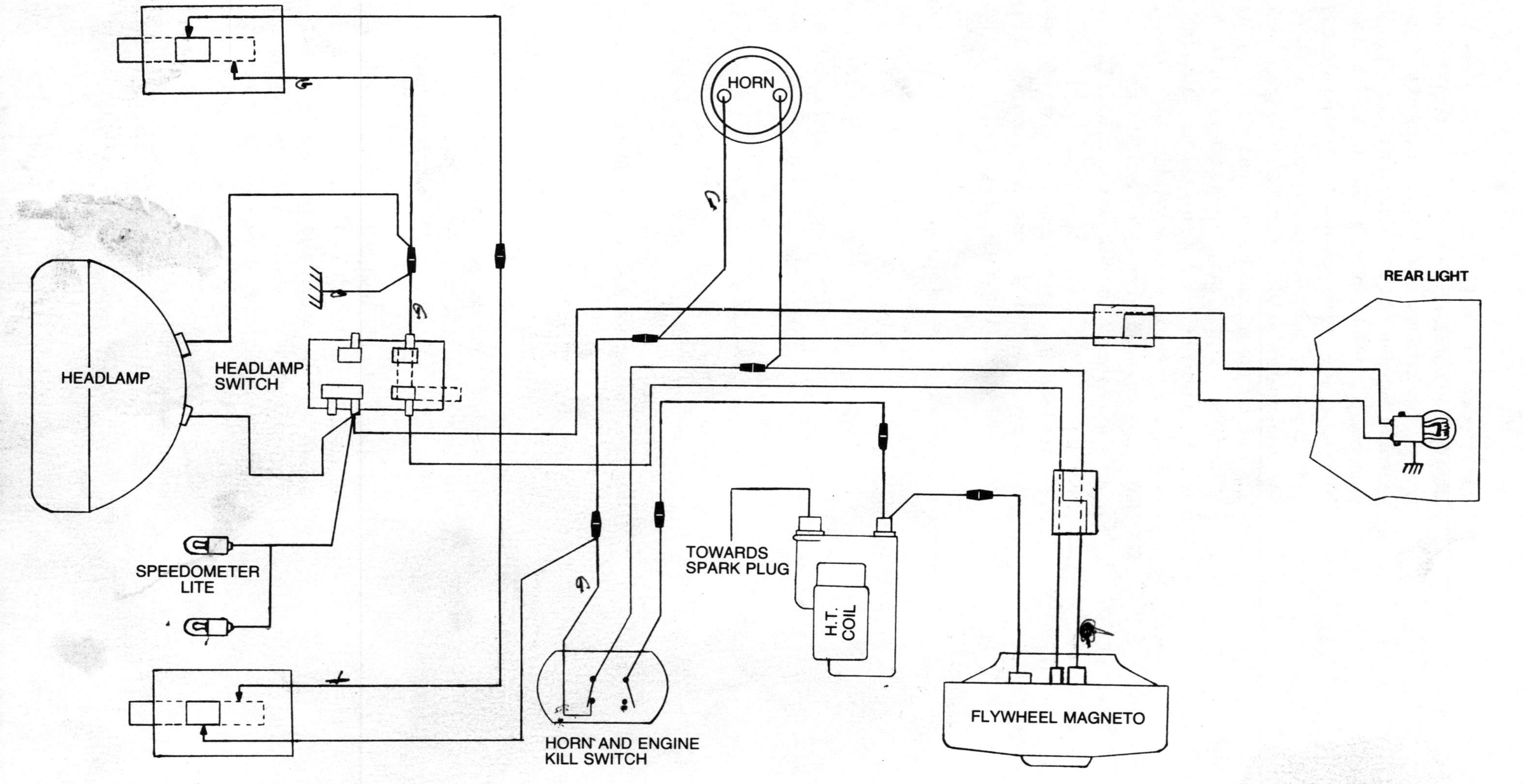

2. The other type of lighting system operates from a battery, so you have lights whether or not the engine is operating, as long as a charge is in the battery. The testing and repair are the same for either system.

If one lamp is out and the others work, then usually the only problem is that the bulb or light unit is burned out. If you have a good bulb or lamp unit, use it to replace the suspected burned-out bulb. If you do not have a new bulb, check the circuit with a test lamp. Remove the bulb or light unit. With the light switch turned on, place one test light prod on a good ground. Place the other prod on the bulb contact. If the test lamp lights, the circuit is good. Then replace the lamp bulb. If the light does not come on, check the circuit for broken wires or loose connections.

If the circuit and bulb are good, make sure the lamp is properly grounded. The bulb is usually grounded through the lamp assembly. Check for rust and a good contact where the lamp assembly is attached to the machine. Make sure the bulb holder is making a good ground to the lamp assembly. If the switch is suspected, bypass it by attaching the wire from the battery or alternator directly to the lighting circuit. A dim light can be the result of a poor ground.

12 MECHANICAL STARTERS

RECOIL STARTERS
WINDUP STARTERS
KICK STARTERS
SERVICING STARTERS

Engines not equipped with an electric starting motor must have some other means of turning the engine crankshaft to start the engine. So various types of mechanical starters are found on different engines.

The simplest method of starting, which is no longer used, consisted of a pulley with a flat surface on the outside, attached to the crankshaft. With a notch in the rim of the pully, a rope about the thickness of clothes line rope and about two and a half feet long was used to crank the engine. To the outer end of the rope was attached a small wooden handle, like a short piece of broom handle or dowel rod with a hole in the middle. A knot held this handle onto the end of the rope, and another knot was inserted on the outside of the notch in the pulley rim. The rope was wound in layers around the surface of the pulley. To start the engine, you grasped the handle and pulled the rope, causing it to unwind and turn the pulley on the engine. While this starting arrangement is no longer used, the basic principle is incorporated into most of today's mechanical starters. Some engines still have a pulley with a notch in the rim so a rope can be used should the other starting system fail.

The starting mechanism is usually mounted on the blower housing assembly, so the unit can be removed without disturbing the starting mechanism. The unit normally requires very little in the way of service. With time and use the rope may wear, fray, or break, or the recoil spring may break.

A certain amount of care when using the rewind (recoil) starter helps to prolong the life of the mechanism:

1. Make sure the screen in the housing, if one is used, is kept clean. Cooling air for the engine moves through the screen and/or filter.

2. Do not allow the starting rope to snap back into the housing. Let it rewind slowly. Releasing the rope when it is fully extended tends to shorten the life of the spring.

3. Do not jerk the rope or pull it all the way out. A smooth steady pull starts the engine under normal conditions.

4. Always pull the starter handle straight out so the rope does not rub against the guide. This practice helps to prevent excessive wear.

With the following description of the different types of starters and the accompanying illustrations, you should be able to disassemble your own unit and replace parts as necessary. There are basically three different types of starter mechanisms.

1. The *recoil* (*rewind*) type, sometimes called a "retractable starter," is the most commonly used. It may be of the "friction shoe" or "cam dog type," this terminology refers to the method of locking the starter mechanism to the crankshaft. A coil spring is used to retract the starting rope. When used on a vertical engine, the construction of this type of starter may be slightly modified. The housing may vary on the snowmobile starter to keep snow and ice from clogging the starter mechanism.

2. The second type is the *wind-up starter*, which may also be referred to as a "self-starter," "impulse starter," "speedy starter," or a "ratchet starter." The wind-up starter uses a ratchet and pawl (dog-cam) mechanism to wind up and to hold the force of a heavy spring. Upon release, the spring turns the engine crankshaft for starting.

3. The third type is the *kick starter*, which may be found on mopeds or small motorcycles. The kick starter has a built-in clutch (ratchet arrangement) to protect the operator and engine from kick-back due to compression or misfiring when starting.

RECOIL STARTERS

The recoil starter operates in the following manner (Figure 12-1). The starter rope is wound around a pulley sheave. When the engine is running, there is no contact between the crankshaft hub or flywheel. When the engine is not running and the starter rope is pulled, the pulley sheave is connected to the crankshaft hub or flywheel. The result is that the engine crankshaft is turned as the rope is pulled. This effect is brought about by a ratchet, cam-dog, or pawl arrangement (basically a one-way clutch) that wedges the pulley sheave to the crankshaft hub or flywheel. When released, the ratchet or cam retracts away from the hub and the rope is rewound on the pulley sheave by a coil spring. On some starters the hub is smooth but constructed to receive the edge of friction shoes, which act to lock the assembly together.

Some side-mounted starters used on vertical crankshaft engines differ: When the starter rope is pulled, a gear moves along a shaft to engage a flywheel with gear teeth on the outer rim. The rope, recoil spring, and ratchet arrangement are all the same as on other recoil starters.

FIGURE 12-1
Recoil starter
Courtesy Kohler Co.

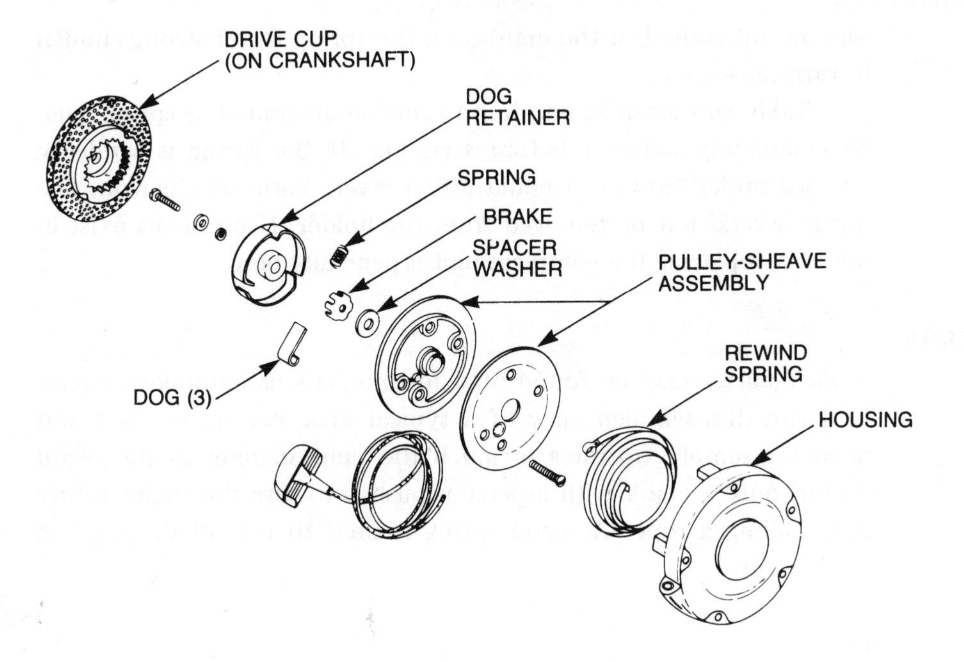

WIND-UP STARTERS

Wind-up starters employ a heavy coil spring, which, when wound up, is used to turn the engine crankshaft for starting. Figure 12-2 shows a disassembled view of a wind-up starter. A handle like a crank is used to wind up a heavy spring. A ratchet and pawl (dog) is used to keep the spring from releasing while being wound up. A reduction gear may also be used on the wind-up handle so it turns easier. When the spring is wound up tight, the engine is ready to be cranked. When the release lever or control is pushed, the ratchet is released and the spring cranks the engine. In cold weather, and particularly if heavy

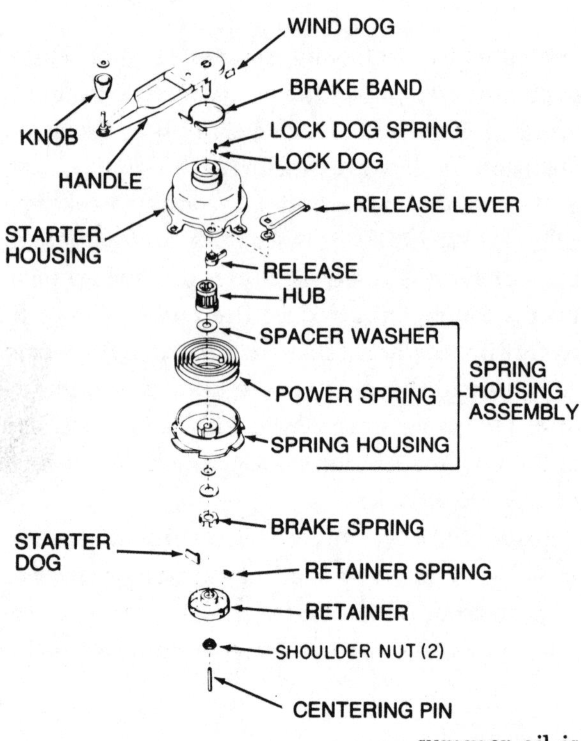

FIGURE 12-2
Wind-up starter
Courtesy Tecumseh Products Co.

summer oil is used in the crankcase, the spring is not strong enough to turn the engine.

Take care when attempting to service the unit. The spring must be completely released before servicing. If the spring is suddenly released under tension, it could inflict severe harm on a person. The spring should not be removed from the holder. If problems exist in the unit, replacing the complete unit is generally best.

KICK STARTERS

A kick starter may be found on a few mopeds or minibikes. Figure 12-3 is a disassembled view of a typical kick starter. A crank and ratchet assembly operates in much the same manner as the recoil starter, only a crank with a pedal is pushed to turn the engine rather than pulling a rope. A recoil spring is used to rewind the starting mechanism.

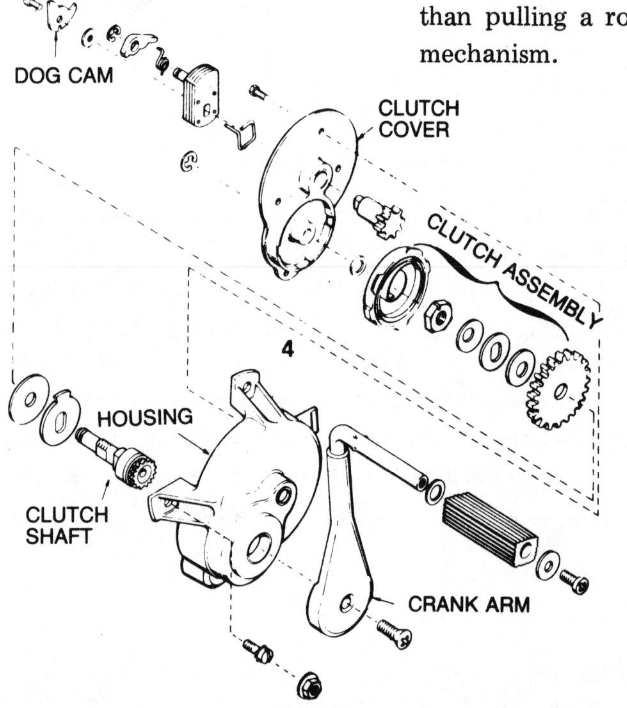

FIGURE 12-3
Kick starter
Courtesy Tecumseh Products Co.

SERVICING STARTERS

Most units do not require any maintenance. If not abused, they function for a long time without problems.

1. Some of the problems, however, that you may encounter with the *recoil starter* include the failure of the rope to rewind. This problem could result from a bent spring, a broken spring, a disengaged spring, a broken rope, binding with the unit, or not enough tension on the spring.

If the unit is noisy during operation, the hub could be rubbing on the cup due to too much end-play.

If the starter is frozen and does not turn, the spring could be broken and jammed on the hub. The cam assembly could be dirty and gummed up, so it cannot operate.

Should the rope or spring break, or should the cam assembly fail to operate, the unit can be removed from the engine and disassembled. To service it, remove the blower housing from the engine. In many cases, the starter assembly is attached to the blower housing. On some engines it is just bolted to the housing, so it can be removed without removing the housing.

The service information is general in nature: After removing the starter assembly from the engine, you should be able to locate the problem. If the spring is broken, it can be replaced. If the rope is broken or frayed, it can be replaced. If there is evidence of wear or broken parts, replacing the entire cam assembly is generally best.

When the rope is to be replaced, cut the knot at the handle and pull the rope out. If there is tension on the rope, pull the rope out part of the way and secure the pulley so it does not rewind. You can do so by clamping the pulley with a vise-grip pliers. Figure 12-4 shows the location of the internal parts of the recoil starter.

To replace the spring, you may have to cut the knot and remove the rope. If there is tension on the spring, remove the handle from

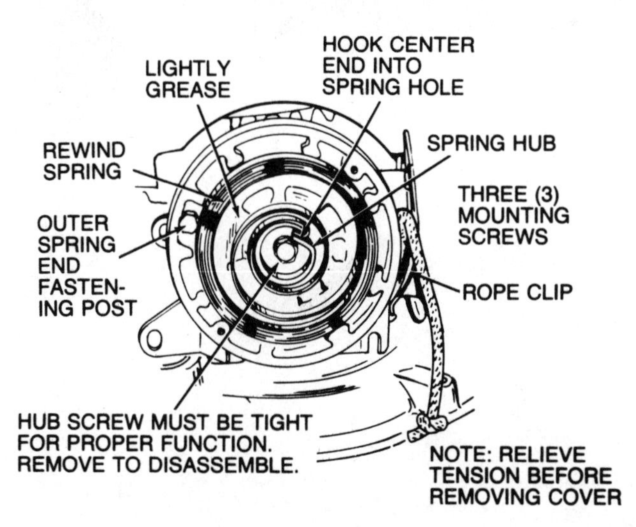

FIGURE 12-4
Recoil starter assembly
Courtesy Tecumseh Products Co.

the rope and let it turn as the spring uncoils. After installing a new spring, you can rewind it by pulling the rope several times.

2. When servicing the *wind-up starter* be sure all the tension is removed from the spring by placing the control in the start position. To check the spring, turn the cranking handle about 10 times. Place the control in "start". If the engine does not turn over, the spring is broken or the ratchet assembly is not functioning.

The spring is normally under considerable tension and can be dangerous unless released carefully. If the spring is wound tight, release the tension by placing the control in "start." Hold the crank handle and remove the screw from the center of it, thus releasing the spring tension. The spring and keeper assembly consistute one unit, so do not remove the spring from the keeper. Some wind-up starters cannot be serviced. The entire unit must be replaced.

3. To service the *kick starter*, remove the crank arm and housing. When removing the housing, hold the dog cam in such a manner that the spring tension is not released. After the housing is removed, gradually let the dog cam turn and release. Figure 12-3 shows a disassembled kick starter. When assembling one, hold the cover with the gear assembly installed and wind the dog cam assembly 2 or 3 turns to apply tension on the spring. Guide the housing into position with tension on the dog cam.

13

COUPLING DEVICES - BELTS CHAINS, CLUTCHES

BELT DRIVES
CHAIN DRIVES
FLEXIBLE COUPLINGS
FRICTION CLUTCHES
VARIABLE DRIVE SHEAVES (PULLEYS)
CENTRIFUGAL CLUTCHES
COMBINATION DRIVE SYSTEMS
INSPECTION, ADJUSTMENT, AND SERVICE OF COUPLING DEVICES
BELT REPLACEMENT AND ADJUSTMENT
CHAIN ADJUSTMENT
FRICTION CLUTCH ADJUSTMENT

To propel a piece of equipment or to drive a particular unit, there usually must be some means of connecting and disconnecting the engine from the driven equipment. For example, if an engine is connected to some type of mechanism that has to be turned while you crank the engine, starting becomes all the more difficult. If the engine is used to propel a piece of equipment from which it is not disengaged, then the entire machine has to move as the engine is being cranked. An exception to this rule would be a hand-pushed rotary mower, whose cutter blades are attached directly to the engine crankshaft.

Many different types of coupling devices are designed to meet the specific requirements of certain pieces of equipment:

1. a simple belt-tightening mechanism between pulleys,
2. a chain located between two sprockets with a tightening device,

3. a friction disc squeezed by spring tension between a driving disc and a driven disc, a combination friction clutch and brake assembly, or

4. a variable coupling that, in addition to engaging the engine to the driven mechanism, provides a variable drive ratio.

Some units are designed to provide permanent torque multiplication by using pulleys of a different size, or through the use of differently sized sprockets.

BELT DRIVES

The simplest coupling device uses a belt idler located between one pulley on the crankshaft and another on the transmission or other unit to be connected to the engine (Figure 13-1). When the driven pulley is to be disengaged from the engine, the idler pulley is released and there is no tension on the belt. When released, the engine drive pulley turns freely without driving the belt. When the idler pulley puts tension on the belt, the engine pulley drives the belt, along with whatever is attached to the driven pulley. In actual operation the belt idler pulley keeps tension on the pulley at all times by spring tension on the idler pulley arm or bracket. To disengage the drive belt on a tractor or riding mower, depress the clutch pedal to overcome idler pulley tension. Many riding mowers use a belt-and-pulley arrangement to drive a twin blade mower unit. The idler pulley is engaged and disengaged by a lever on top of the mower housing.

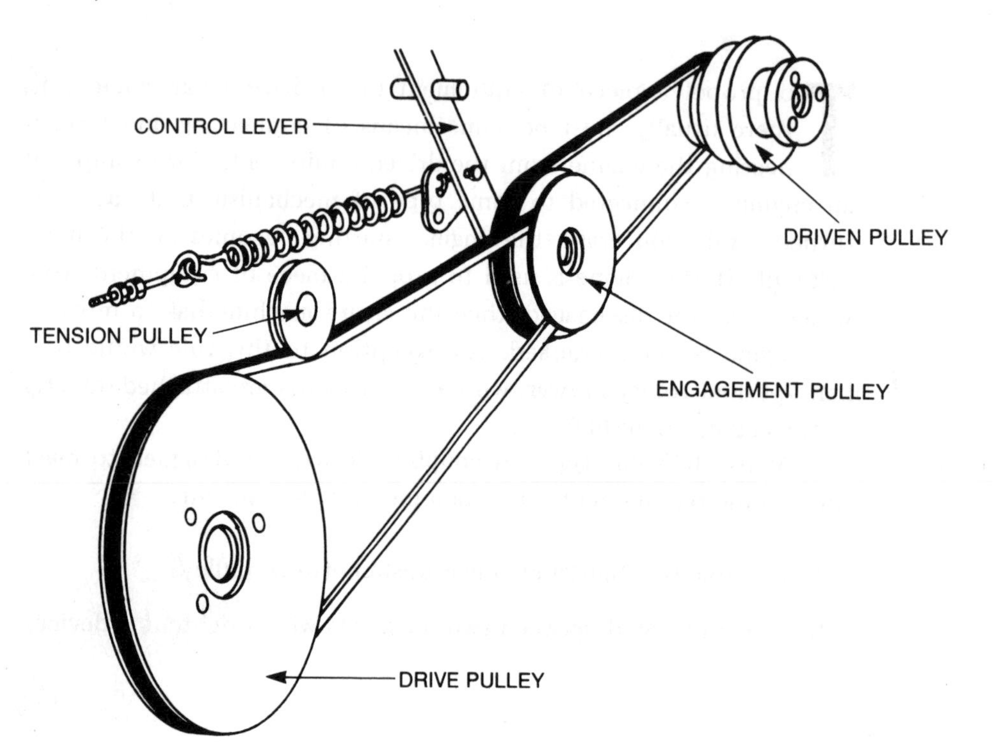

FIGURE 13-1
Drive belt engagement pulley
Courtesy Deere & Company, Moline, Illinois

CHAIN DRIVES

On some machines, particularly mopeds, sprockets are used either in place of pulleys or along with belt-driven units. The size of the sprockets can vary to give whatever ratio is needed. A sprocket is constructed much like a gear with widely spaced large teeth. Sprockets and a chain, for example, comprise the standard drive on bicycles.

Although a chain, which is used to connect the sprockets, acts the same as a belt, it is more stable than a belt. It does not stretch or fray, and, if wear does take place, the chain can be shortened by removing links. A chain may be used in the open, or it may run in oil enclosed in a housing. Figure 13-2 shows the use of a chain and sprocket on a moped. Provisions are made on a moped, as on a bicycle, for moving the rear wheel assembly to take up chain slack.

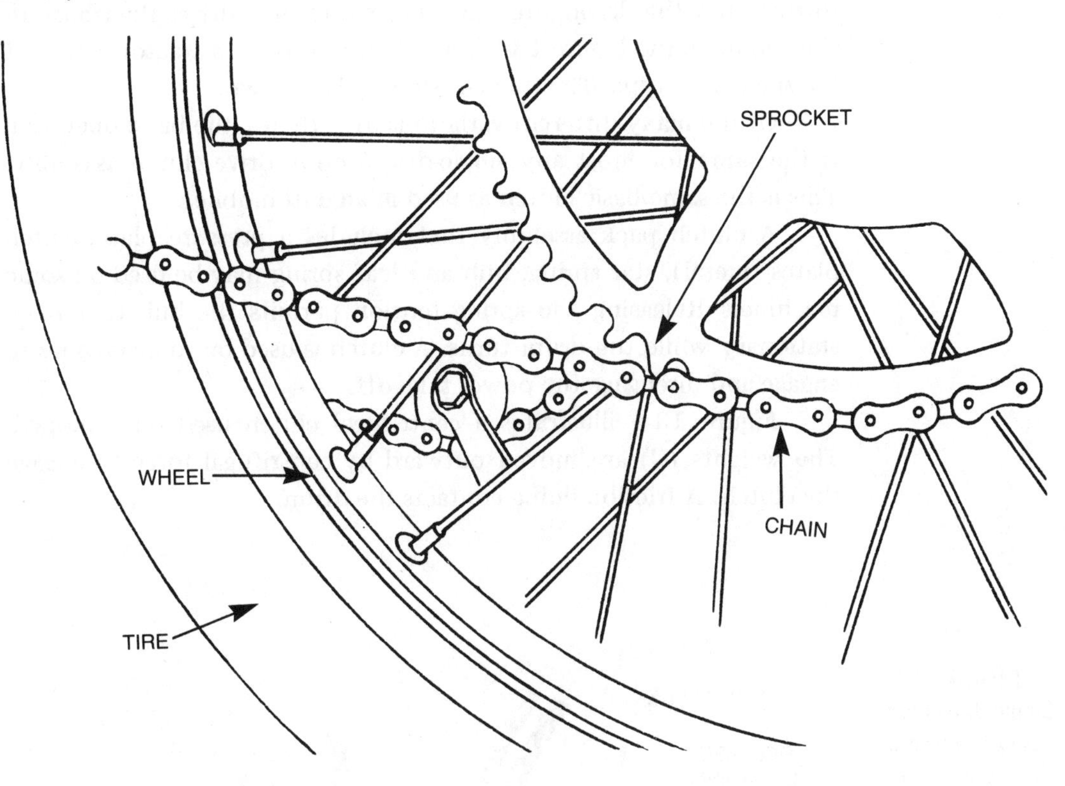

FIGURE 13-2
Chain and sprocket
Courtesy Motobecane America, Ltd.

FLEXIBLE COUPLINGS

A power unit, such as a tractor or riding mower, may use a long shaft to connect the engine to the driving unit. To prevent vibration and to compensate for flexing between the drive and the driven units, a flexible coupling may be used at the rear of the shaft. One yoke is attached to the drive shaft, and another similar yoke is attached to the shaft to be driven. A flexible ring is located between the two yokes, which are bolted to the flex ring opposite one another. The shaft torque is carried by the flex ring, which is usually made of a heavy fiber like material.

FRICTION CLUTCHES

A friction disc clutch is used on some tractors and mowers. The purpose is to engage and to disengage the engine from the driven units. On this type of clutch, a drive plate is attached to or driven by the engine crankshaft. A driving disc, which usually has an asbestos facing riveted onto it, is attached to the transmission input shaft. On either side of the disc is a pressure plate assembly. When the clutch is applied, that is, when the clutch pedal released, a heavy spring or set of springs applies pressure on the pressure plates, which lock the drive plate to the driving disc. The result is that the engine turns the transmission input shaft. A release mechanism, connected through a linkage to a pedal, permits the operator to release the spring tension by depressing the clutch pedal. The pressure plate assembly continues to turn, but the driving disc no longer turns and drives the transmission input shaft. Figure 13-3 is a cut-away view of a tractor clutch of the single-disc type. The transmission is also shown.

Of the many different variations of clutches, the basic operation is the same for most any single-disc friction drive clutch assembly. This is the same basic clutch as used in an automobile.

A clutch pack assembly that includes a pressure plate, clutch plates (metal), star spring, hub and leaf spring may be used on some machines. Releasing the spring tension permits the hub to remain stationary while the drum turns. A clutch is used on some engines to engage and disengage the power take-off.

Figure 13-4 illustrates a centrifugal clutch used on a moped. The weights (C) are moved outward by centrifugal force to engage the clutch. A friction lining contacts the drum.

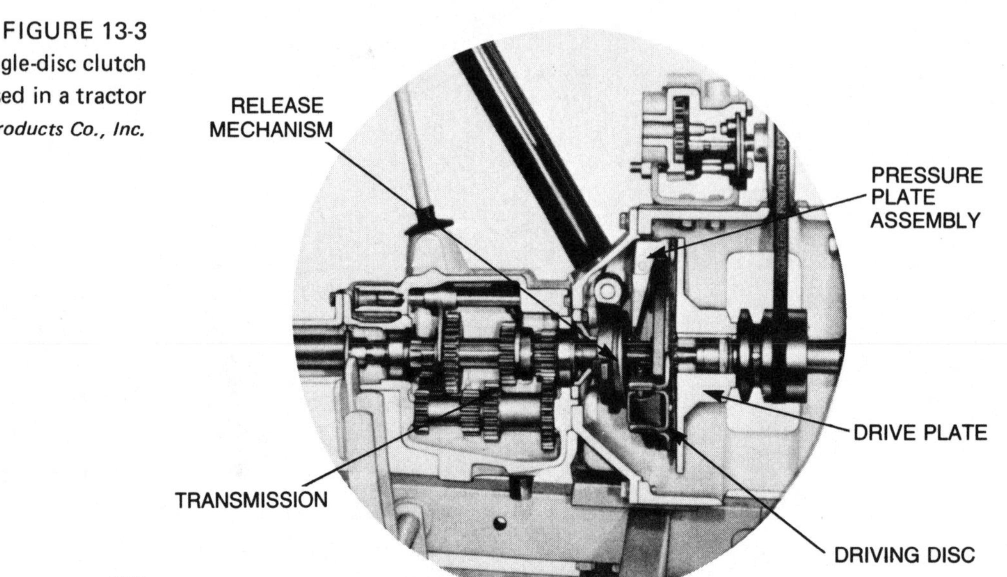

FIGURE 13-3
Single-disc clutch used in a tractor
Courtesy Engineering Products Co., Inc.

FIGURE 13-4
Centrifugal clutch using linings
Courtesy Motobecane America, Ltd.

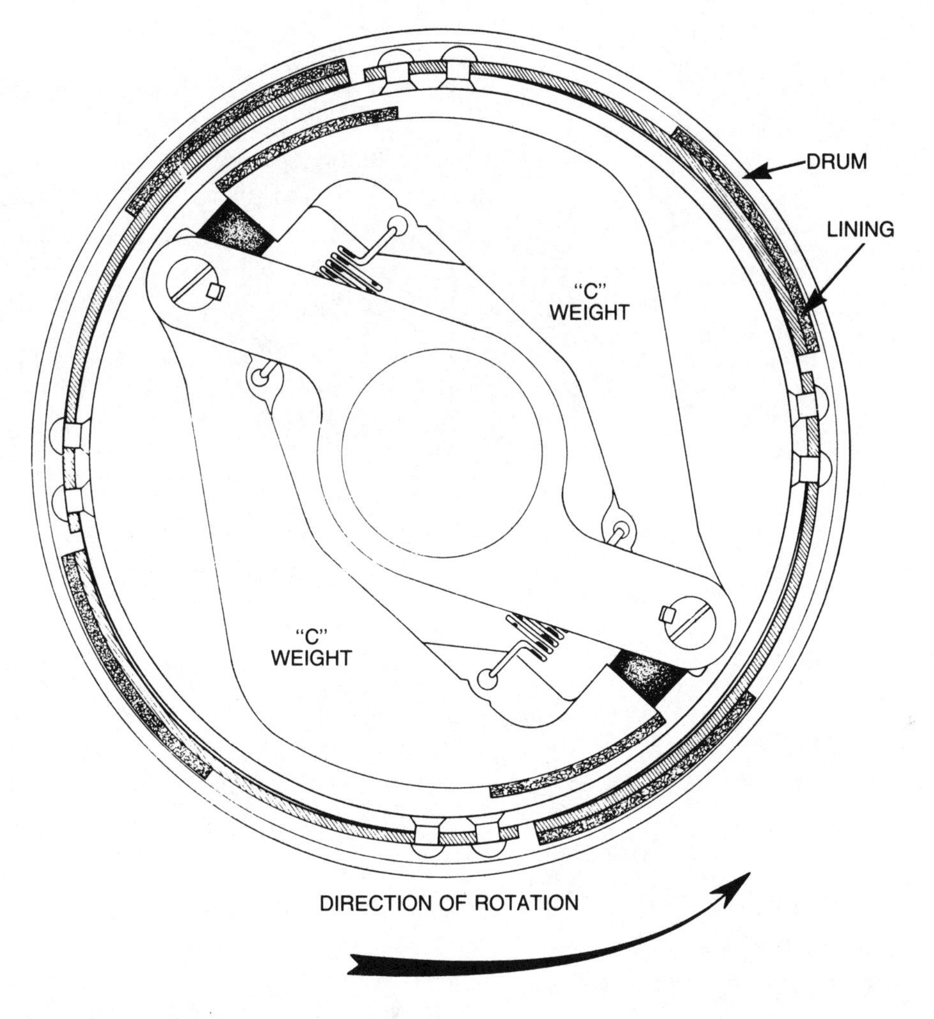

VARIABLE DRIVE SHEAVES (PULLEYS)

A number of machines use a variable drive arrangement that acts as a clutch to disengage the engine from the driven unit at idle. It also automatically changes the drive ratio by varying the width of the flanges of the driving pulley, upon which the drive belt operates. Basically it involves a movable flange on the crankshaft pulley, which moves closer to the stationary flange causing the belt to ride higher in the pulley. The results are less reduction and increased speed.

On some installations only one pulley reacts, and the engine has flexible mounts so it can tilt to keep tension on the belt. Other installations use two pulleys having variable pulley flanges.

A number of different designs are used to obtain the same results, namely to provide for disengagement at idle, full power at low speeds, and low power at high speeds. Essentially this "variable ratio transmission" comes under different trade names used by different manufacturers, but the basic operation is very similar.

Many mopeds use a single-variable drive sheave to power the chain drive unit (Figure 13-5), and the drive sheave is attached to the engine crankshaft.

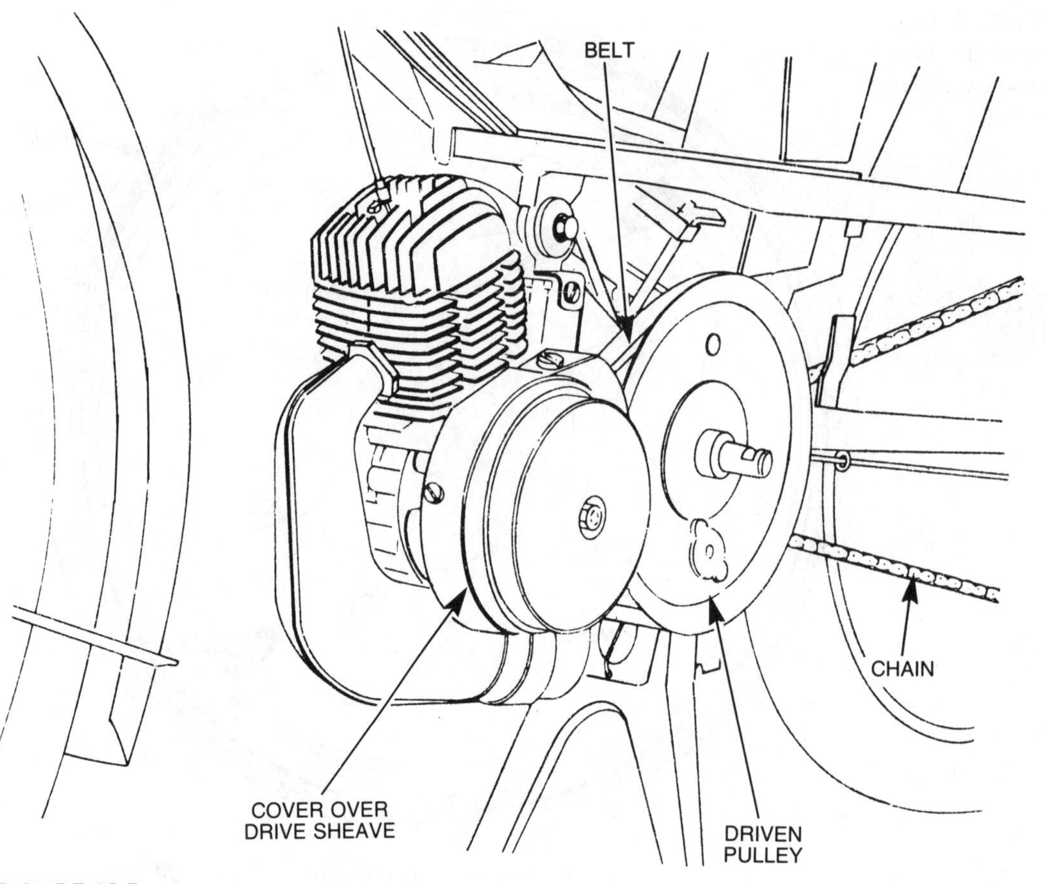

FIGURE 13-5
Single-variable drive sheave
Courtesy Motobecane America, Ltd.

A variable drive sheave is basically a pulley to drive a vee belt. As the pulley's one movable half (face) move toward the stationary side of the pulley, the belt is forced to move up in the pulley. This movement causes the driven pulley to turn faster. As the pulley faces spread apart or move together, the diameter of the pulley varies, thus modifying the ratio to better serve the load imposed on the engine.

The movable face of the pulley is controlled by centrifugal force acting on weights inside the pulley, which take the form either of balls or other weights and which can move outward under speed. To compensate for the variation in belt length, the engine is attached to flexible mounts so it can tilt to maintain the necessary tension on the belt. Figure 13-6 shows a flexible engine mount.

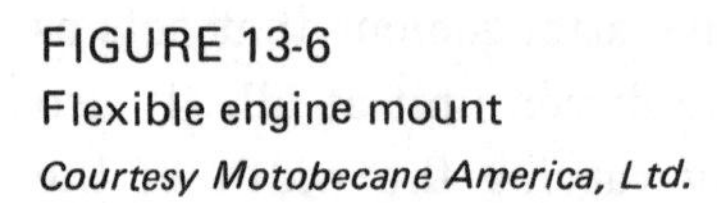

FIGURE 13-6
Flexible engine mount
Courtesy Motobecane America, Ltd.

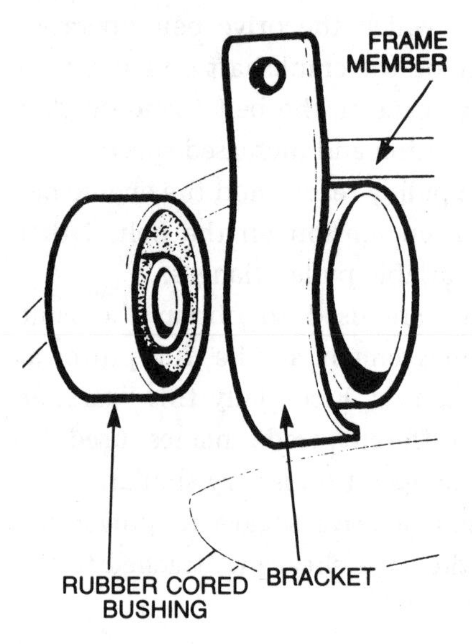

When starting at low revolutions, the pulley is spread apart and the belt remains stationary on the smallest pulley diameter. As soon as the engine begins to pick up speed, the engine, which is held forward by spring tension, begins to drive the pulley with the belt on the smallest diameter. The reduction ratio is at its greatest, considered low range. As engine speed increases, the balls or weights between the two halves of the pulley begin to move outward due to centrifugal force. As they do so, the movable face of the pulley is moved toward the stationary pulley face (Figure 13-7). This effect

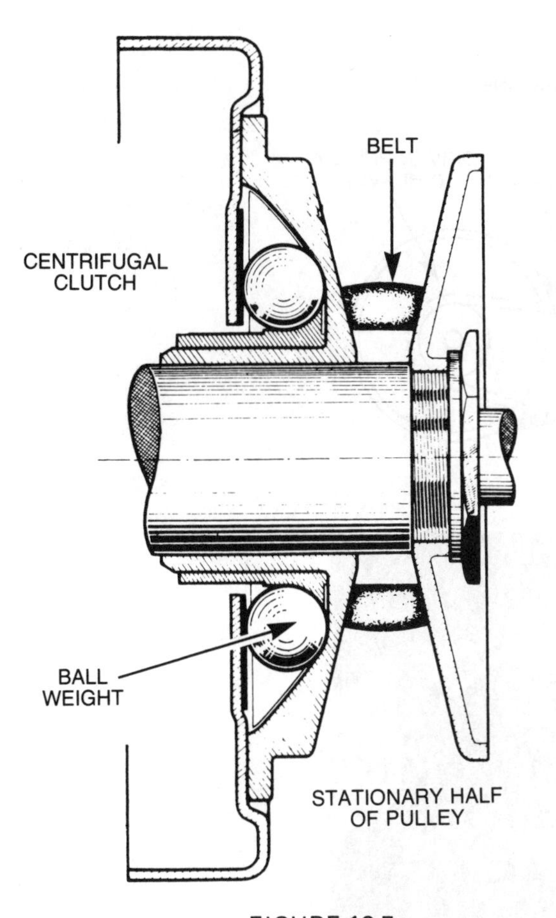

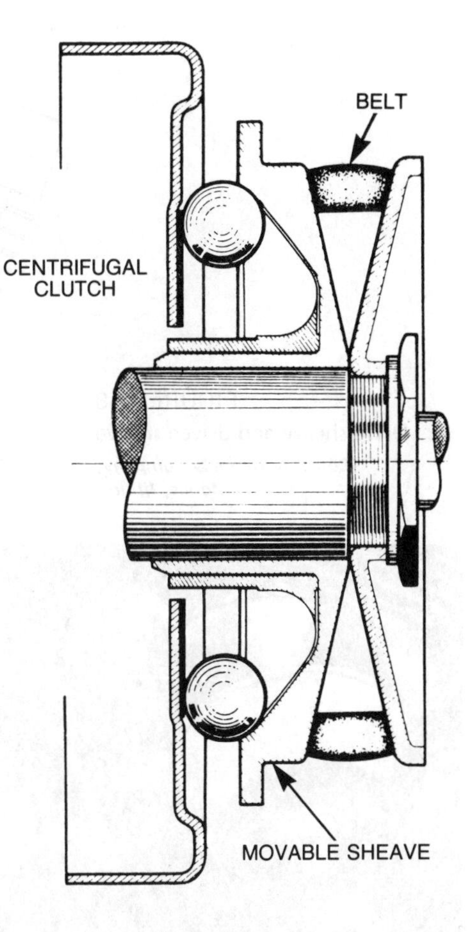

FIGURE 13-7
Operating diagram of a variable-sheave pulley
Courtesy Motobecane America, Ltd.

causes the belt to move up toward the rim of the pulley. In other words the belt operating diameter of the pulley is increased until top speed is obtained. When a hill is encountered or some other type of load that tends to slow the machine, the weights drop down. In so doing, they make the driving face of the pulley smaller and thus modify the reduction ratio. Basically this type of drive reacts like an automatic transmission. With speed change there is a continuous modification of the reduction ratio.

A variation of this type of drive is found on some golf carts and snowmobiles, where flexible motor mounts would be impractical. The drive sheave (pulley) operates in the same manner. When the engine reaches top speed, the sheave halves are as close together as possible. The belt rides out toward the outer top face of the sheaves. The driven sheave is constructed so that one pulley face is stationary and the other half is movable, with the two halves held together by spring action (Figure 13-8). This arrangement not only acts as a take-up for the action of the drive sheave, but it is also torque sensitive.

Figure 13-9 shows the drive sheave and the spring-loaded driven sheave. The driven sheave rides on a cam bracket as it opens (faces spread apart) to obtain the high-speed position. The turning force keeps the driven sheave in the low-speed position. When an increased load is encountered with the machine at top speed, such as when

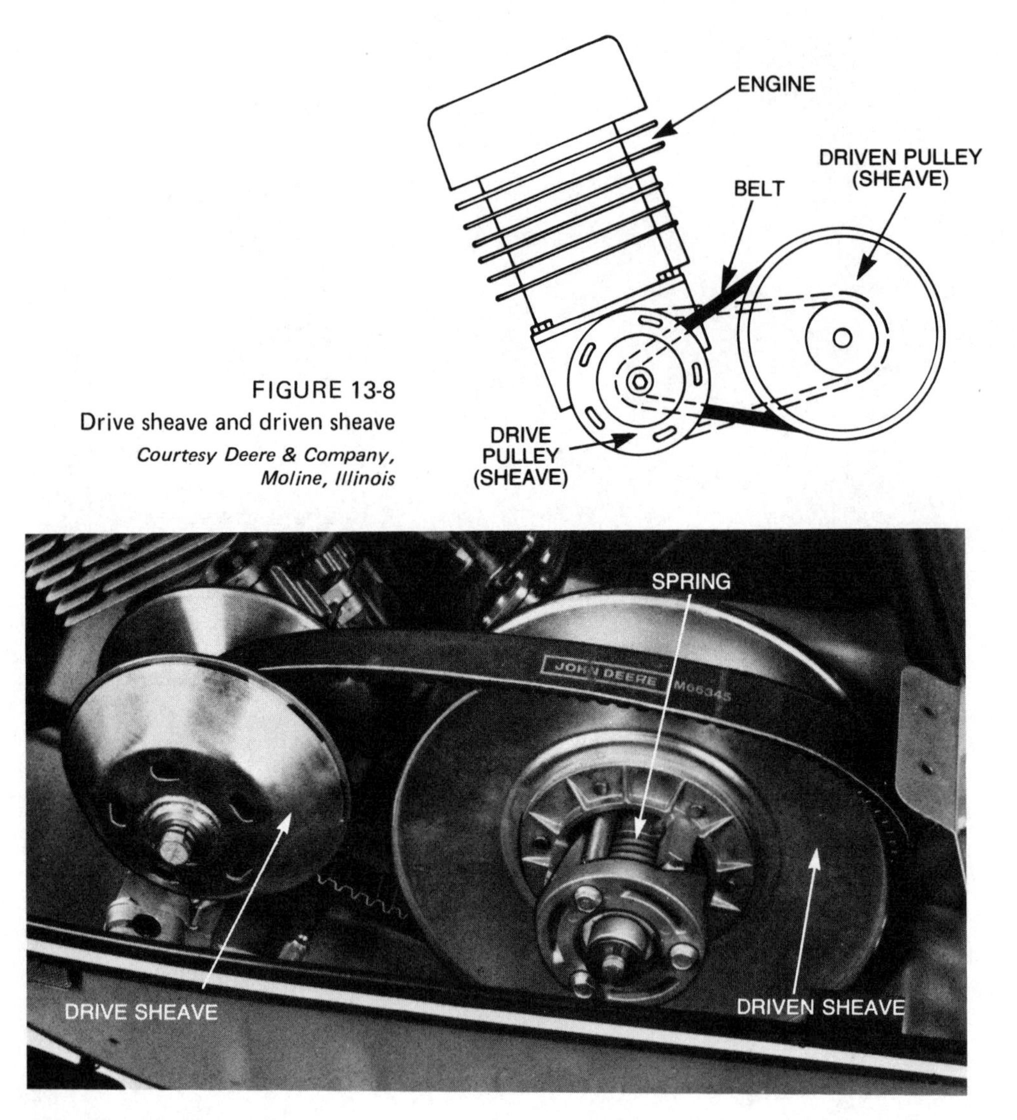

FIGURE 13-8
Drive sheave and driven sheave
Courtesy Deere & Company, Moline, Illinois

FIGURE 13-9
Spring-loaded driven sheave
Courtesy Deere and Company, Moline, Illinois

climbing a hill, the cam bracket in the driven sheave forces the two halves of the pulley together, thereby obtaining a slower travel speed while maintaining high engine revolutions for increased torque.

CENTRIFUGAL CLUTCHES

Some units, commonly chain saws, use centrifugal clutches to connect the driven units to the engine when a certain speed is obtained. The engine must be turning at speeds of between 3,000 to 4,000 rpm before the cutting chain begins to revolve. The unit shown in Figure 13-10 basically consists of three clutch shoes. The drum and chain sprocket are separate units, and a gear to drive the chain oil pump is mounted on the hub. When the engine is at idle the clutch shoes are held in the idle position by spring tension. As engine speed increases centrifugal force overcomes spring tension and forces the clutch

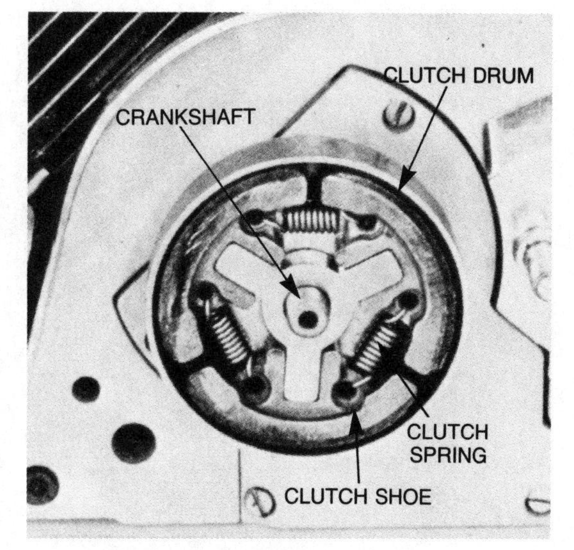

FIGURE 13-10
Centrifugal clutch using three shoes
Courtesy Stihl Corporation

shoes outward against the clutch drum, thus transmitting engine torque to the chain drive sprocket.

Another type of centrifugal clutch is used on a moped for smooth engagement and disengagement of the final drive (Figure 13-4). As the assembly is revolved, flyweights in the form of levers that are pivoted on one end move outward. The free end of the weights exert pressure on the revolving drum. As speed increases the drum and hub assemblies are locked. As speed is reduced spring tension pulls the weights away from the drum mechanism and the hub assembly no longer drives the drum.

COMBINATION DRIVE SYSTEMS

The drive systems on many installations may be combinations of units. Nomenclature can sometimes be confusing, however, because various manufacturers may label their units differently. For example, the pulley, which has one half fixed solidly on a shaft and the other half mounted so it can slide in or out to change the width of the vee that the belt rides in, may carry different names. The units may be called "variable pulleys," "variators," "clutches," "drive sheaves," or "driven sheaves." Yet the units, while not constructed in exactly the same manner, all perform the same function. They serve as a clutch as well as a torque converter to provide varying ratios according to load requirements.

A number of drive systems may use a combination of units to obtain the best final drive operation. The main elements of some snowmobile drive and coupling systems may include a drive sheave, a drive belt, a driven sheave, a drive sprocket, a drive chain, a driven sprocket, and a drive shaft with track drive sprockets. Figure 13-11 shows the drive shaft and chain sprocket used to drive the track on a snowmobile.

FIGURE 13-11
Track drive shaft and sprocket
Courtesy Deere and Company, Moline, Illinois

The main drive elements of a moped may include a drive sheave, a belt, a driven pulley, drive shaft, drive sprocket, chain, and driven sprocket. Figure 13-5 shows the elements of a moped drive and coupling system.

A centrifugal clutch may also be incorporated into a system to provide positive disengagement and engagement of the drive and driven units. The characteristic of a centrifugal coupling is that the faster it is turned, the more pressure it applies to hold the units in engagement.

INSPECTION, ADJUSTMENT AND SERVICE OF COUPLING DEVICES

Many coupling devices, such as belts and chains, are adjustable. The friction clutch release linkage is adjustable on most installations. The variable drive sheaves normally provide maintenance-free service over a long period. When malfunctions occur, it is generally best to replace the entire unit. In most cases the couplings are located in such a position that you can easily check their condition and operation. Belts and chains can readily be checked, and in most cases you can watch the operation of the variable sheave coupling to see if it moves in and out according to load and speed.

BELT REPLACEMENT AND ADJUSTMENT

A belt that is cracked, frayed, or stretched should be replaced. Belt tension is important: Too little tension results in belt slippage, while too much may cause the belt to stretch and create excessive wear on the components that the belt drives. A belt that squeals upon acceleration indicates slippage.

Belt tension can be easily checked on straight drive belts. Place a straight edge across the pulleys, and then exert thumb pressure on the belt midway between the pulleys. A properly adjusted belt should deflect 1/4 inch with normal thumb pressure. Generally a slotted bracket, which helps to hold the unit in place, permits you to move the unit in or out to adjust the belt after loosening the attaching bolts.

Belts used to drive mower blades are generally engaged by using a belt idler which is controlled by a lever. The idler pulley is typically spring-loaded and keeps uniform pressure on the belt. If the belt slips, you can usually move the mower housing to take up the surplus slack. Belt guides are used to keep the belt from slipping out of the pulley when the idler is released. Some manufacturers specify the belt be adjusted until there is a specific distance between the inside face of the belt at a particular place on the housing. The manufacturer's instructions give the specifications and procedure for making such an adjustment.

Figure 13-12 shows a belt arrangement used to propel a tractor or riding mower, as well as for operating mower cutting blades or other equipment. A lever is used to engage the equipment, and a pedal is used to release the drive belt. A spring is used to maintain tension on the idler pulley. An arrangement similar to this is used on a great many riding mowers and lawn tractors.

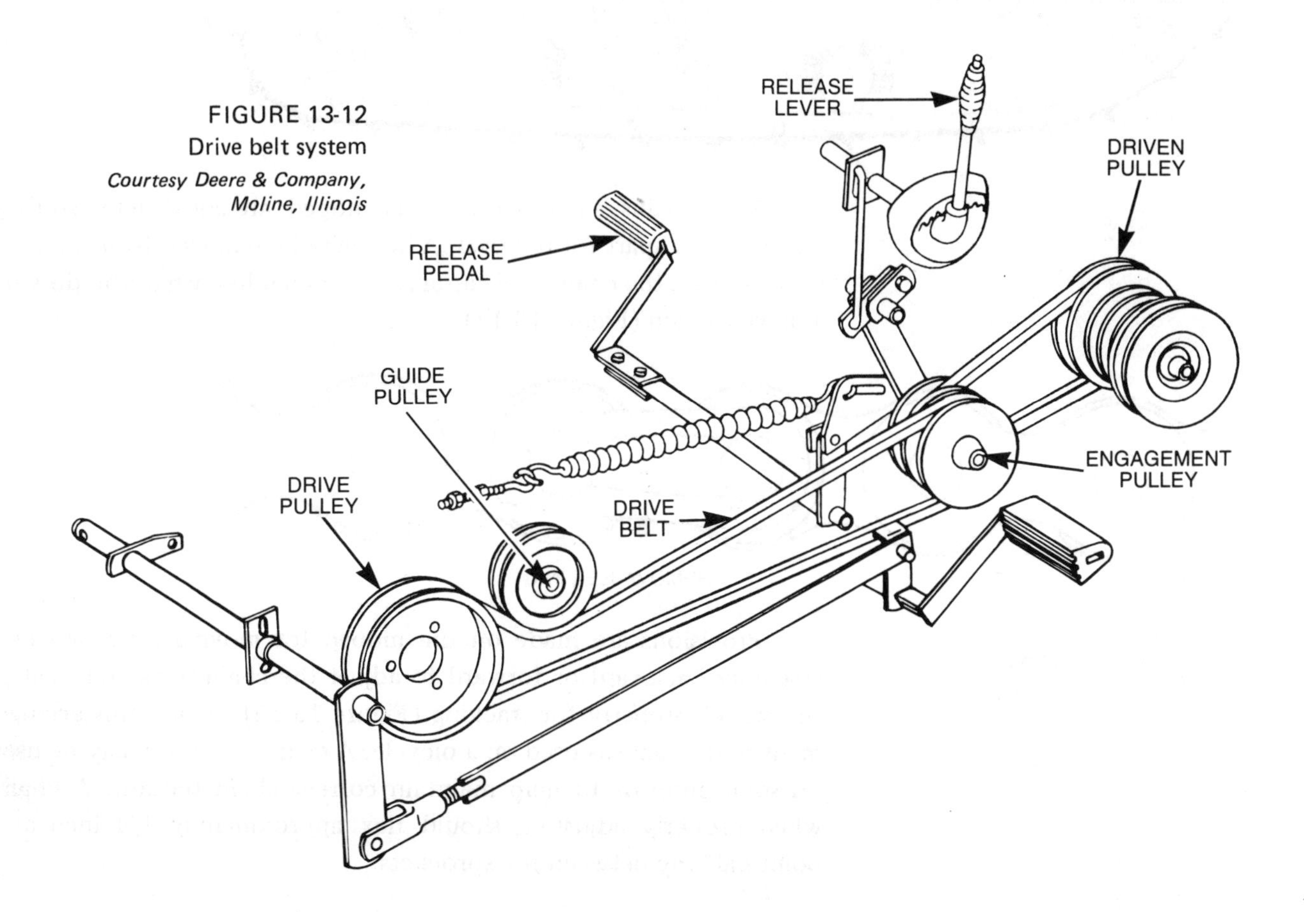

FIGURE 13-12
Drive belt system
Courtesy Deere & Company, Moline, Illinois

When replacing or removing a belt, be careful not to stretch it. With a variable sheave pulley, try rotating the movable half of the pulley and working the belt down in the groove to obtain slack (Figure 13-9).

When a belt idler is used to provide belt tension, the pulley must run true and in alignment with the belt. If the bushing and/or shaft, which the pulley runs on, is worn so the pulley wobbles, then either the shaft, the bushing, or both should be replaced. Spring tension on the idler pulley arm is usually adjustable.

CHAIN ADJUSTMENT

Some installations have provisions for adjusting the drive chain tension. Figure 13-13 shows the chain tension adjustment used for the drive chain of a snowmobile. The track on a snowmobile is treated in the same manner as a drive chain. The chain is adjusted so that there is 1/4-inch free play or deflection in the center. The rear of the machine must be supported off the ground.

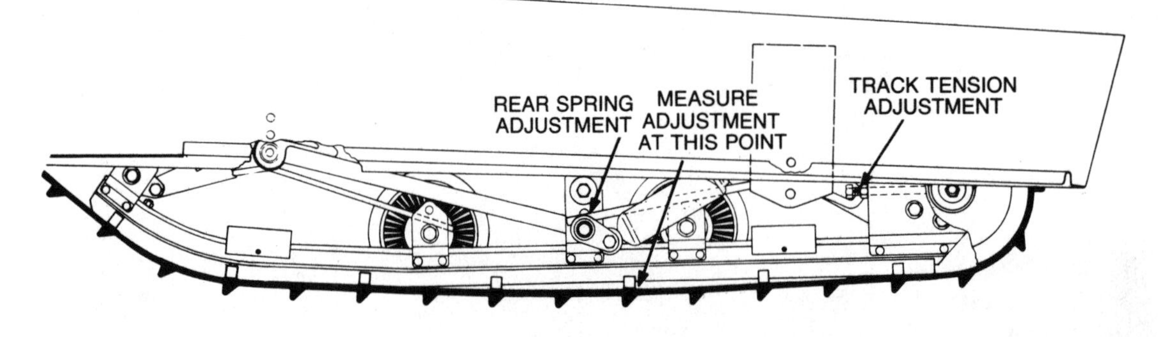

FIGURE 13-13
Snowmobile track adjustment
Courtesy Polaris E-Z-Go Division of Textron Inc.

Most chains, such as those on a moped, are constructed so they can be taken apart for removal. A link may be removed from some so as to shorten the chain if necessary, very much like what you do with a bicycle chain (Figure 13-14).

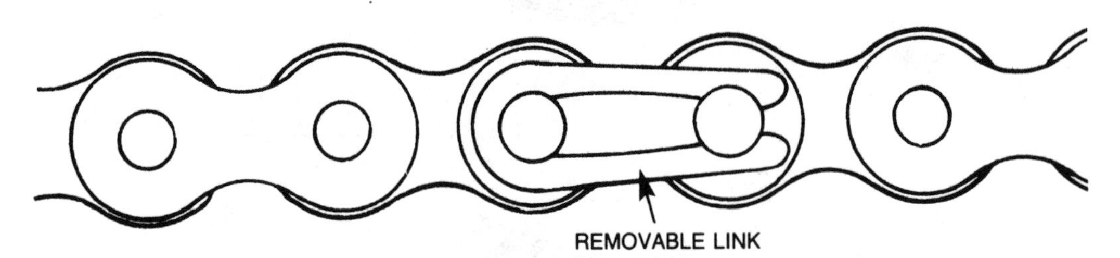

FIGURE 13-14
Removable chain link
Courtesy Motobecane America, Ltd.

Provisions are made on the moped for moving the rear wheel assembly backward or forward to adjust the chain slack and to align the wheel properly for tracking (Figure 13-15). Again, this arrangement is the same as used on a bicycle. A chain tensioner may be used on some mopeds to help maintain correct chain tension. A chain, when properly adjusted, should flex approximately 1/4 inch at a point halfway between the sprockets.

FIGURE 13-15
Adjusting rear wheel and chain
Courtesy Motobecane America, Ltd.

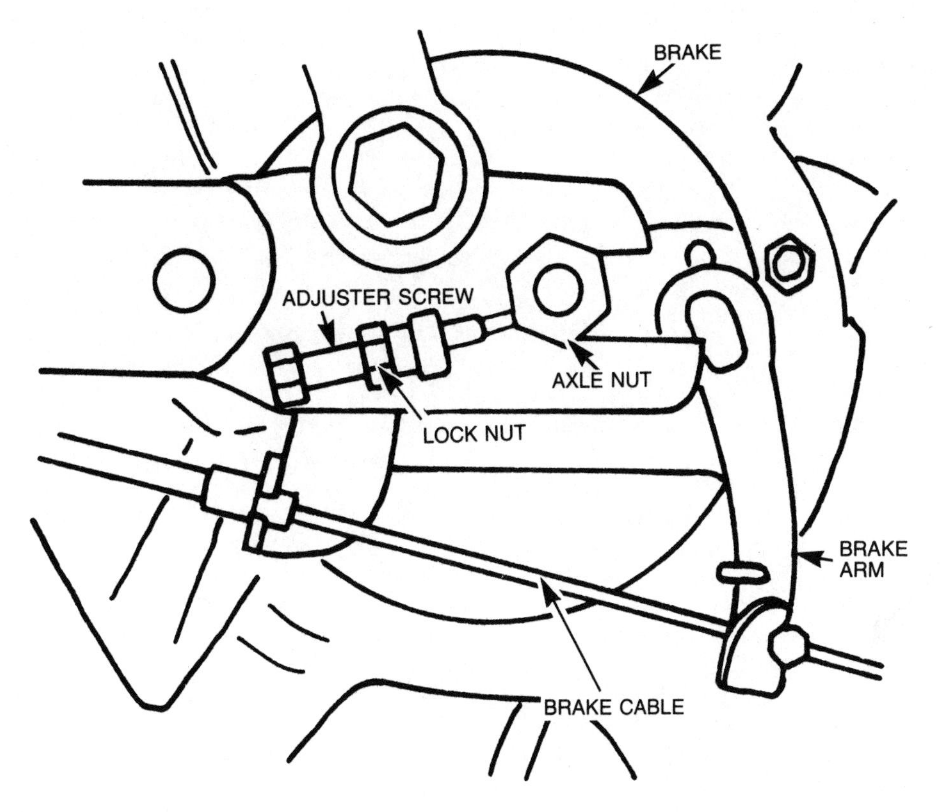

FRICTION CLUTCH ADJUSTMENT

The only adjustment on most friction disc clutch assemblies is the pedal linkage. Some installations, such as a power take-off clutch, may use a lever to release the clutch. In any case, the linkage is used to control engagement and release.

The linkage must be adjusted so there is approximately 3/8 to 1/2 inch of pedal or lever free-play, which is the movement before you feel the release bearing contact the pressure plate release fingers. The adjustment usually entails lengthening or shortening the linkage. If the clutch still slips, grabs, or does not release properly with correct linkage adjustment, then the clutch unit must be removed from the machine. It is generally best to replace the entire clutch assembly, including the disc, pressure plate assembly, and the release bearing.

14

SUSPENSION SYSTEM, STEERING, WHEELS, TIRES

SUSPENSION SYSTEMS
SNOWMOBILE FRONT SUSPENSION
SNOWMOBILE REAR SUSPENSION
GOLF CAR SUSPENSION SYSTEMS
MOPED SUSPENSION
STEERING SYSTEMS
STEERING FORKS
STEERING GEARS AND STEERING LINKAGE – FOUR-WHEEL MACHINES
SNOWMOBILE STEERING
WHEELS
MOPED WHEELS
TRACTOR WHEELS
TIRES
CHECKING SUSPENSION SYSTEMS
CHECKING STEERING SYSTEMS
CHECKING WHEEL ALIGNMENT
TIRE REMOVAL

Suspension systems are generally used on machines intended to transport people. Such systems are designed to prevent road irregularities from being transmitted to the machine and its passengers. They do so through the use of some type of spring or cushioning system, which may take the form of coil springs, leaf springs, torsion bars, or shock absorbers. Shock absorbers are generally used to complement the suspension system. Some type of suspension system is used on snowmobiles, golf carts, and mopeds. Lawn and garden tractors and riding mowers normally do not use any type of flexible suspension system other than, in some cases, a spring-type seat or cushion.

A machine that transports people must also give operators the means to change the machine's direction—or in other words, to guide or steer it. With a single wheel in front, directional control is a simple matter. Steering can be done directly with handle bars which control a fork, such as on a moped, or with a tiller arrangement, as used on some golf carts. In place of a tiller, many of the newer golf carts, which use one front wheel, may have a regular steering gear mechanism to control the direction of the wheel.

With two wheels—or with two skis, as in the case of the snowmobile—the wheels or skis are connected together so they are both controlled by the same steering mechanism. The linkage is so designed, however, that the inside wheel or ski turns at a sharper angle when rounding a corner. This sharper angle gives stability by preventing the wheels from sliding and also makes for better control.

When a steering wheel is used, some sort of gear mechanism must convert the wheel's rotary motion into back-and-forth movement. It should also provide a gear reduction for easier steering and help to absorb road shock when operating on an irregular surface.

Wheels provide the rolling surface on which the machine travels. On self-propelled machines with four wheels, two are the driving wheels, generally the rear wheels. When only two wheels are used, such as on a moped, the back wheel is the driving wheel. The snowmobile is supported on an endless track or belt, which is also used to propel the machine.

Wheels are supported directly or indirectly on antifriction bearings of the ball or roller type. This arrangement keeps friction and drag at a minimum while supporting a load, and in the case of the driving wheels helps to propel the machine.

Rubber tires are used on most machines to provide traction and to cushion shock from the surface that the machine is traveling on. Heavy machines and machines used to carry people generally have pneumatic tires, which are tires filled with air and which provide flexibility for better traction and operator comfort. Solid rubber tires are used on most hand-type mowers, whether pushed or self-propelled. The solid rubber tire requires no maintenance and offers some cushioning and traction benefits over a steel wheel. Many snowmobiles use solid rubber tires on the wheels that support the track.

SUSPENSION SYSTEMS

The suspension system is designed to be compatible with the machine it is used on. The snowmobile probably has the most complex suspension system. Despite variations among the different makes and models, all such systems attempt to obtain the best ride for the specific machine.

SNOWMOBILE FRONT SUSPENSION

The following information constitutes a general overview of some of the common front suspension systems. While different components may be used, the basics are all much the same. Figure 14-1 shows a typical snowmobile front and rear suspension system.

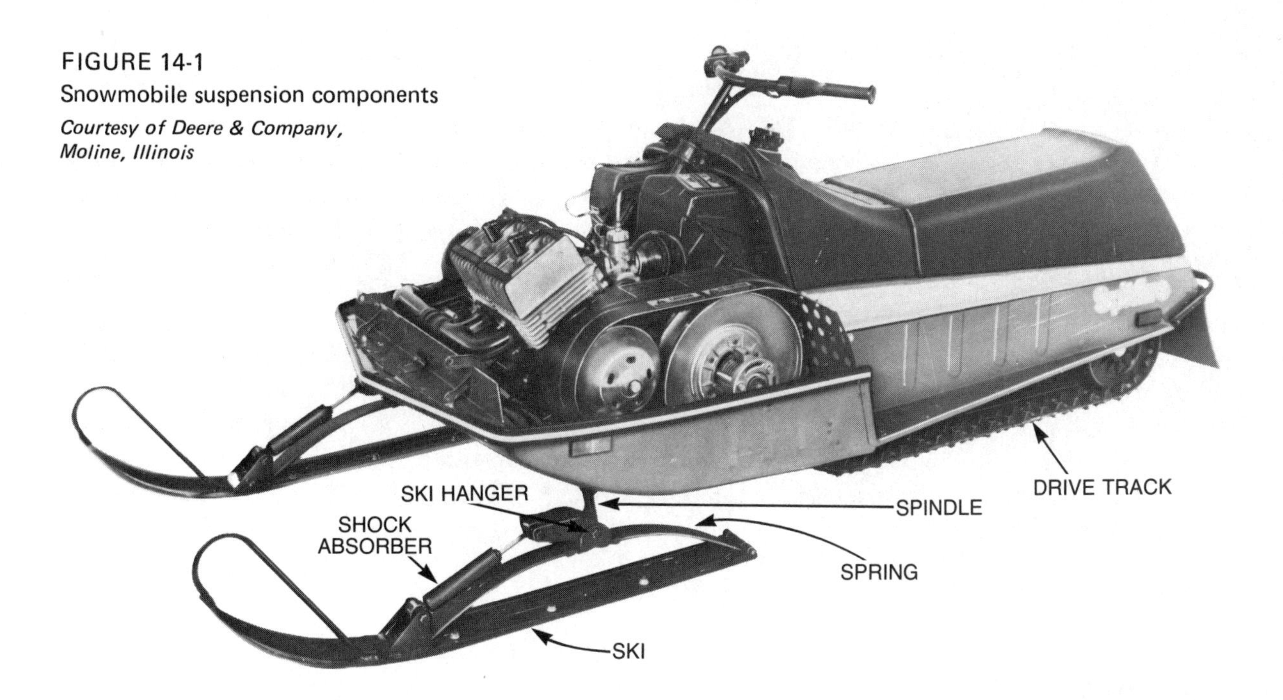

FIGURE 14-1
Snowmobile suspension components
Courtesy of Deere & Company, Moline, Illinois

A pair of skis are used to support the weight of the front end of the machine. The skis, which are mounted on spindles so they can be turned for steering purposes, take the place and serve the same purpose as the front wheels on a lawn or garden tractor. The skis have replaceable wear plates for prolonged service. The suspension system is located between the skis and the body to cushion the roughness of the terrain for the passengers. In addition the steering system is part of the suspension system.

Of all the suspension arrangements on the various machines, some use combinations of the different components. The simplest front suspension system is a leaf spring mounted near each end of the ski. The spring is attached to the snowmobile chassis about one-third of the distance back from the center of the spring. The spring is attached in such a manner that one end can move back and forth to take care of the lengthening and shortening of the spring as it flexes. Figure 14-1 shows a typical spring.

The leaf spring arrangement is generally used in conjunction with a shock absorber for greater stability. The shock absorber complements the spring as well as checking rebound. When a spring deflects it tends to continue bouncing until all the energy is absorbed. When a shock absorber is used in conjunction with the spring, the shock absorber tends to check this rebound. As shown in Figure 14-1, the shock absorber on this installation is located at an angle, which adds stability to the ski. The hanger, which attaches the ski to the chassis, acts as a pivot so that each end of the ski can move up and down to follow the contour of the terrain.

Another type of front suspension system uses a coil spring over the outside of the shock absorber. When the machine hits a severe bump, the coil spring compresses, thus complementing the shock absorber. In addition each ski is attached through linkage to a torsion bar. A torsion bar is a steel bar to which the ski linkage is attached in such a manner that as one ski goes up the bar twists. This reaction is the same as a spring flexing, but it gives added stability if downward pressure is applied on one ski as the opposite ski is forced upward. Spring action is obtained as the bar twists and untwists.

Suspension systems must be carefully designed. Too soft (flexible) a suspension system results in too much rebound, which can be controlled to some extent by shock absorbers. Too soft a suspension system makes for unstable steering control. Too stiff a suspension system results in a hard ride.

SNOWMOBILE REAR SUSPENSION

The rear suspension of the snowmobile supports the track that drives the machine and also carries a portion of the weight of the machine and the occupants. In most cases the track assembly is adjustable to compensate for rider weight and the varying terrain (the depth of the snow).

The drive track is made of a synthetic rubber like an endless belt or drive chain. Some tracks use a reinforcing material, such as fiberglass. Rubber lugs are molded on the inside of the track to engage the drive sprockets. Various lugs are used on the outside track surface for traction. Steel cleats and bars may be used for better traction. Removable wear bars, used on many tracks, permit the installation of new bars to provide better traction. The tracks are usually 15 inches wide, and they are guided by aluminum slide rails on either side.

Figure 14-2 shows a typical track and suspension unit partially removed, along with the suspension unit after the track has been removed. Coil springs support the load through a shaft arrangement. Idler wheels with rubber tires are mounted on shafts both front and rear to support the track. Shock absorbers are used to cushion the ride. The spring tension can be changed to make for better riding and handling. The track is driven by special rubber sprockets mounted on a drive shaft, which in turn is driven by the engine (refer back to Figure 13-11).

Little wear takes place on the suspension components, except for the units that come in contact with the surface—snow or ground. The removable wear bars and rods on the skis and track can be replaced by simply unbolting the worn parts and bolting on new units in their place.

FIGURE 14-2
Removing snowmobile track and suspension and suspension with track removed
Courtesy Deere and Company, Moline, Illinois

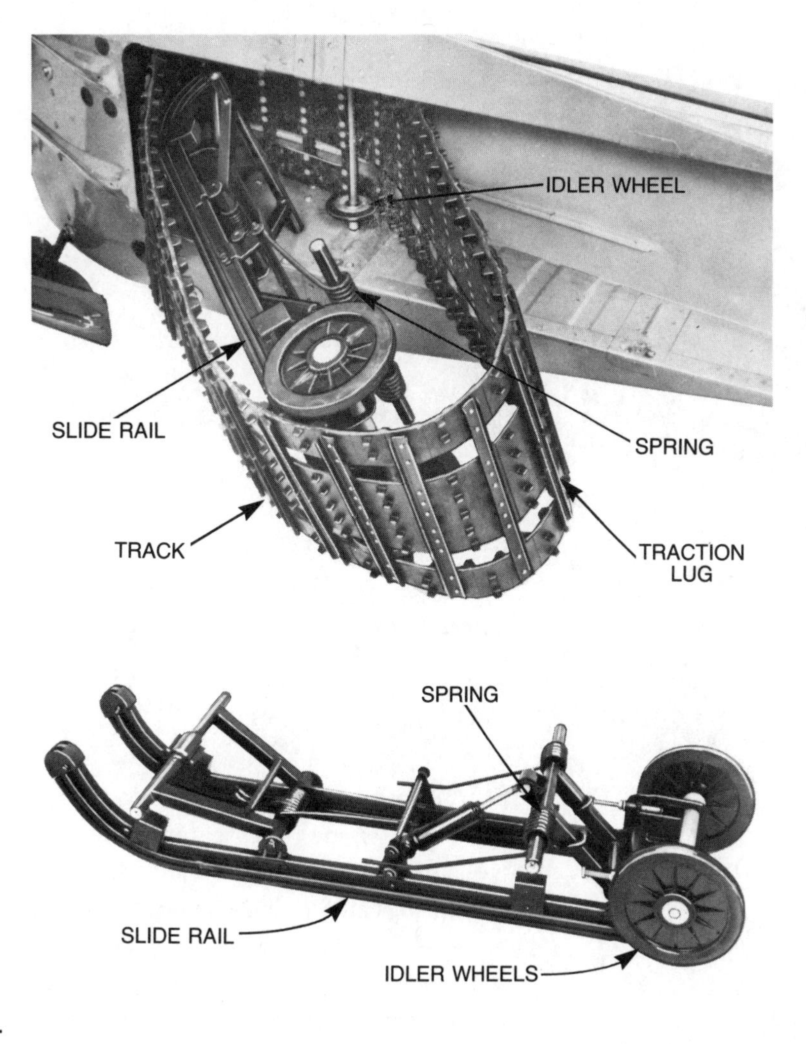

GOLF CART SUSPENSION SYSTEMS

Golf cart suspension systems vary among the different makes and models. Older models may use leaf springs in the rear to carry the weight and cushion the ride. One spring is located on each side parallel to the frame, with each end of the spring attached to the frame. The front end of the spring is attached solidly while the rear end of the spring is attached through a shackle (link) arrangement, to permit the spring to lengthen and shorten as it flexes. Shock absorbers may or may not be used to check spring rebound. The rear axle (the driving axle) is clamped to the center of the spring.

On most of the later model carts, a coil spring is located on either side of the cart between the rear axle housing and the frame. A stabilizer bar may be used to provide fore and aft stability to the driving axle. Shock absorbers may be located inside the two coil springs, or one shock may be located at the center between the frame and the differential housing (Figure 14-3).

Coil springs are generally used at the front on four-wheel carts. The coil springs are used between each suspension arm and the frame. Some carts use a suspension tube to support the weight on the wheel, permitting each wheel to move up and down without affecting the

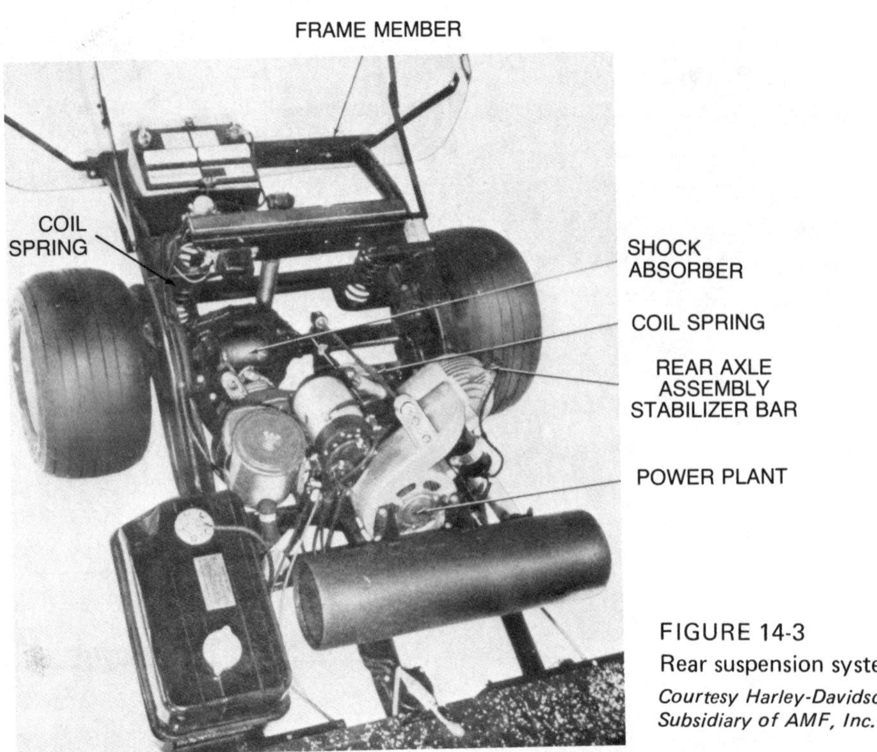

FIGURE 14-3
Rear suspension system for golf cart
Courtesy Harley-Davidson Motor Co., Inc. Subsidiary of AMF, Inc.

opposite wheel. It has much the same effect as a torsion bar. A shock absorber may be located inside each coil spring or between the wheel support and frame. Figure 14-4 shows a four-wheel cart using coil springs and an independent-type suspension.

In place of coil springs, some carts use a leaf spring of the transverse type on the front. The center of the spring is clamped to the center of the frame cross member. The ends of the spring are attached to the wheel support assembly.

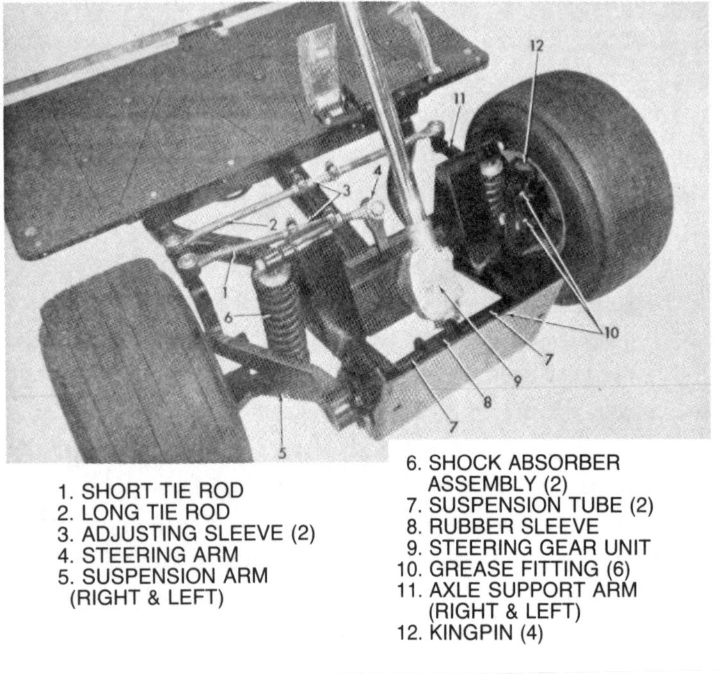

1. SHORT TIE ROD
2. LONG TIE ROD
3. ADJUSTING SLEEVE (2)
4. STEERING ARM
5. SUSPENSION ARM (RIGHT & LEFT)
6. SHOCK ABSORBER ASSEMBLY (2)
7. SUSPENSION TUBE (2)
8. RUBBER SLEEVE
9. STEERING GEAR UNIT
10. GREASE FITTING (6)
11. AXLE SUPPORT ARM (RIGHT & LEFT)
12. KINGPIN (4)

FIGURE 14-4
Front suspension system for golf cart
Courtesy Harley-Davidson Motor Co., Inc. Subsidiary of AMF Inc.

Three-wheel carts use a fork arrangement to steer and to suspend the front wheel. Common practice is to use a two-piece fork assembly. The fork tubes, to which the wheel spindle is attached, slide inside the section of the fork that is attached to the frame and that can be turned for steering. A coil spring is used between the end of the sliding tube and fork to cushion the road irregularities. Figure 14-5 shows a disassembled view of this type of suspension.

Some three-wheel carts use not springs at the fork, but a linkage arrangement instead. An arm about 6 to 8 inches long is mounted parallel to the ground on each end of the fork. Each arm is attached to a shaft that is mounted into each end of the fork. Like a torsion bar, the shaft twists or turns slightly under pressure. The wheel spindle with the wheel is mounted on the arm at about the midway point. A shock absorber, attached at the end of each arm with the other end fastened to the fork, acts to cushion road bumps and assist the torsion rods.

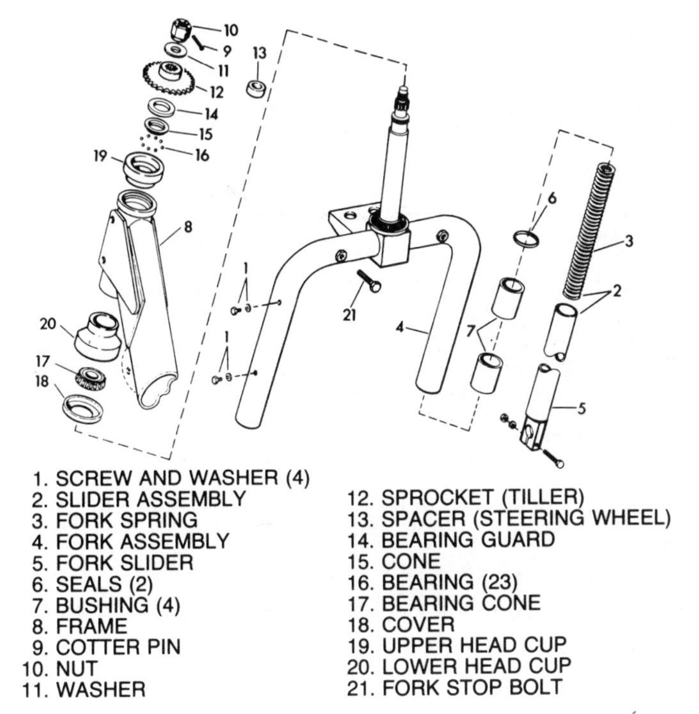

FIGURE 14-5
Front fork steering and suspension unit used with three-wheel golf cart
Courtesy Harley-Davidson Motor Co., Inc. Subsidiary of AMF, Inc.

MOPED SUSPENSION

The front suspension used on a moped is very similar to the arrangement used on the three-wheel golf cart. The lower section of the fork slides into the upper fork assembly, an arrangement that is known as a "telescoping fork." Coil springs located inside the fork act to cushion the road shock as the wheel moves up and down in the fork. Figure 14-6 shows a typical front fork used on a moped.

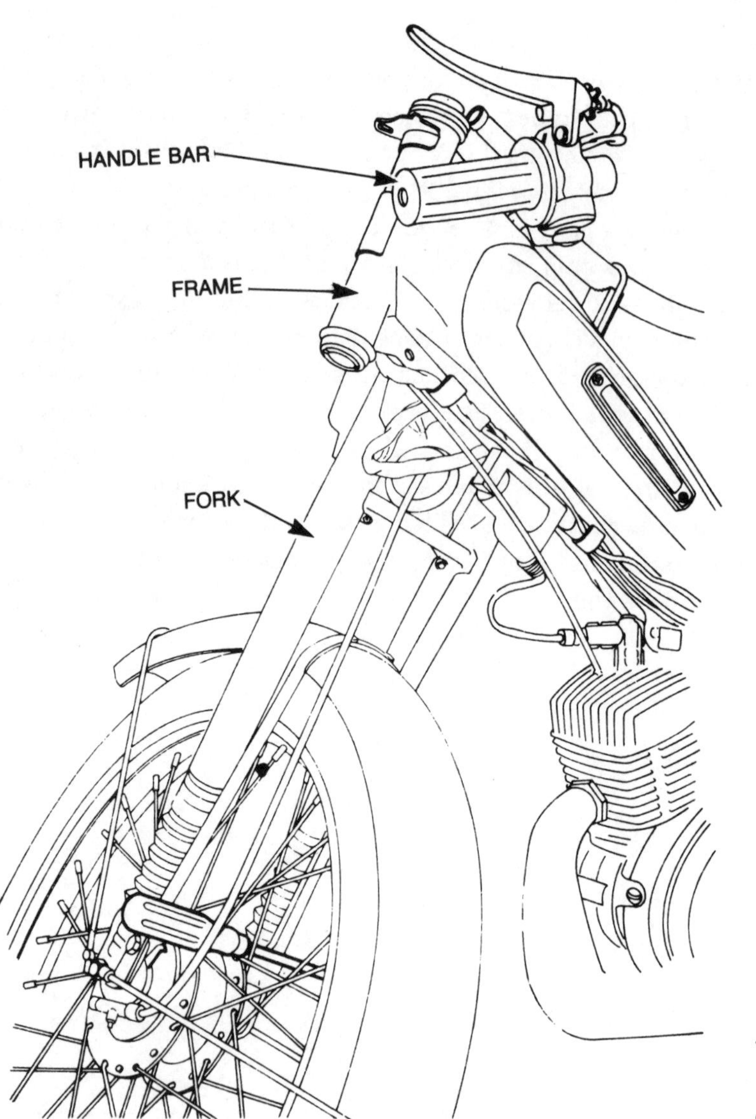

FIGURE 14-6
Typical moped front suspension removed from frame
Courtesy Motobecane America, Ltd.

Most mopeds use a swinging arm suspension at the rear; that is, the rear wheel assembly is attached by a swing arm to the lower section of the frame. The swing arm is hinged on the frame mounting bolt. A shock absorber is attached to each side between the upper section of the frame and the end of the swing arm near the wheel spindle (Figure 14-7). Some installations have a coil spring over the shock absorber to assist when severe bumps are encountered.

Despite the variations in suspension systems, their construction is quite simple, making it generally easy to locate troubles and service. Shock absorbers and coil springs are common to most suspension systems. If there is no resistance when bouncing the machine, the shock absorber is probably faulty. If there are signs of oil leaking around the shock, it is probably beginning to fail. Broken springs are very obvious. Weak springs are usually easy to locate.

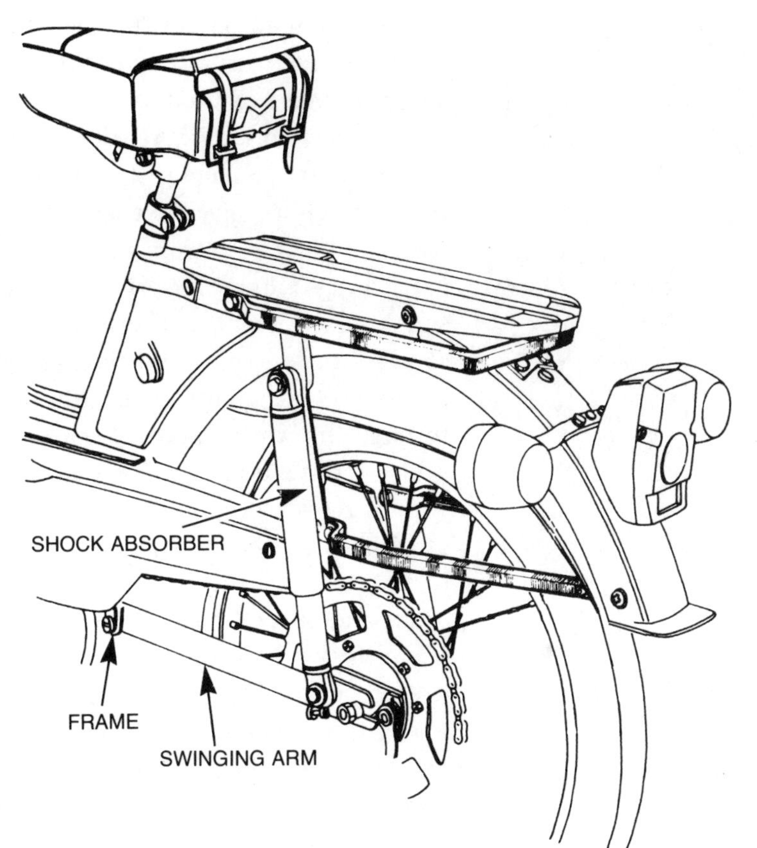

FIGURE 14-7
Typical moped rear suspension
Courtesy Motobecane America, Ltd.

STEERING SYSTEMS

Any machine, propelled either by hand or by power, must enable the operator to guide it in the direction desired. The simplest steering is the swiveling device used on a walk-behind lawn mower, whether it is pushed or self-propelled. A handle of some type, extending behind the mower, permits the operator to swing the machine in whatever direction desired. The front wheels are mounted on swivels, which are offset from the center line of the wheels in the same manner as a furniture caster. Thus these wheels trail when the machine is pushed.

Although the following information is general, it shows the basic operation of typical steering systems, because it applies to machines in common use.

STEERING FORKS

On mopeds and three-wheel golf carts, the front wheels are supported in a fork arrangement, very much like the fork used on a bicycle. A U-shaped tubular frame mounts the wheel, and a center post supports the fork in the frame of the machine. The center support post is mounted on antifriction bearings so it can turn easily. On the moped handlebars are attached directly to the post of the fork so any movement of the handlebar is directly transmitted to the wheel through the fork (Figure 14-6). The fork is always mounted in the frame in such a manner that it creates a trailing effect, which makes for steering stability. It also accounts for the fact that, when you remove your hands from the handlebars, the machine tends to travel in a straight line. To achieve this effect, the frame member that supports the fork is placed at an angle, that is, it is tilted backward at the top. If a center line is extended downward through the center of the fork, it would contact the road surface ahead of a center line through the wheel. This off-center arrangement, known as "caster," brings about a trailing effect on the front wheel.

Three-wheel golf carts may use either a tiller or a steering gear assembly to steer the front wheel. When the tiller is moved from side to side to guide the cart, the movement is transferred to the fork. A chain and sprocket may be used to provide a mechanical advantage between the fork and the tiller bar. It provides more stability and easier steering than if the tiller is connected directly to the fork, Figure 14-8. Shown also are the support bearings that support the fork in the frame. A sliding block, which fits into a tiller guide attached to the fork in place of a chain and sprockets, is used on some of the later models to provide a reduced steering effort.

A regular mechanical steering gear assembly, used on late model three-wheel carts, is connected by linkage to the fork (Figure 14-9). It steers in the same manner as a four-wheel cart.

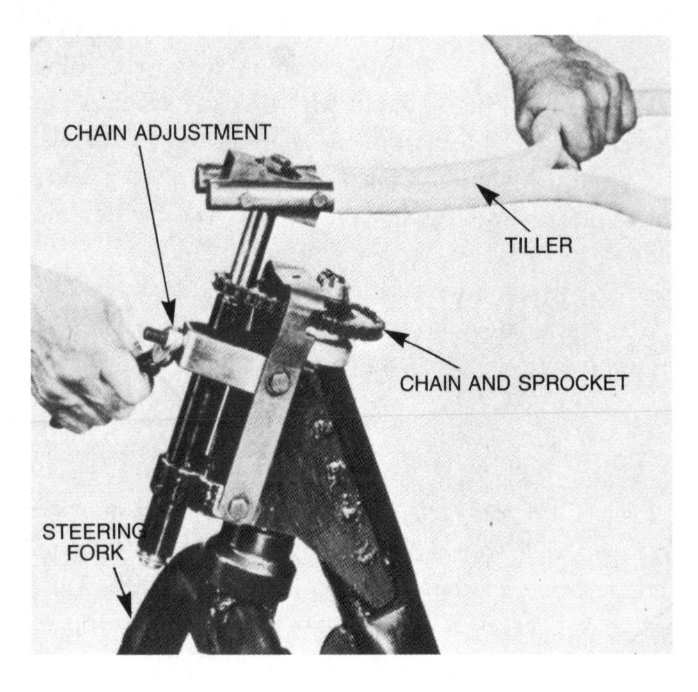

FIGURE 14-8
Tiller steering using chain to turn steering fork
Courtesy Harley-Davidson Motor Co., Inc. Subsidiary of AMF, Inc.

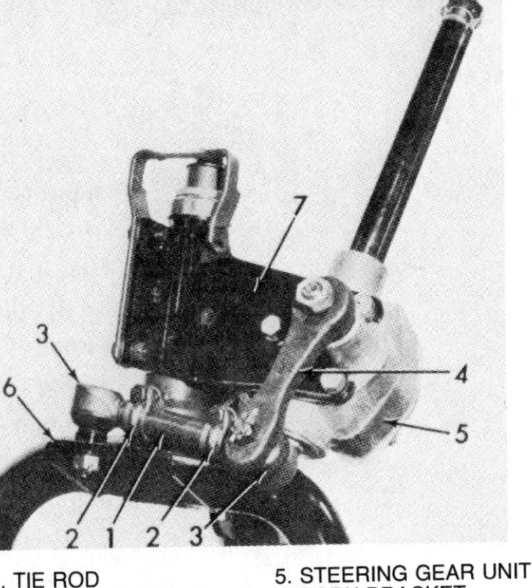

FIGURE 14-9
Steering gear mechanism (automotive type)
Courtesy Harley-Davidson Motor Co., Inc. Subsidiary of AMF, Inc.

STEERING GEARS AND STEERING LINKAGE—FOUR-WHEEL MACHINES

The steering gear used on many golf carts and/or some tractors is the same type used on automobiles. This type of steering converts revolving motion from the steering wheel into back-and-forth motion at the steering linkage arm. The steering wheel is attached to one end of the steering shaft. On the other end of the shaft, a worm gear is in constant mesh with a steering arm shaft. The gear mechanism is mounted on antifriction bearings in the steering gear housing. The steering arm shaft extends through the housing. One end of the steering arm is attached to the steering arm shaft, and the opposite end is attached to the steering linkage. Figure 14-10 is an exploded view of the steering gear showing all the parts.

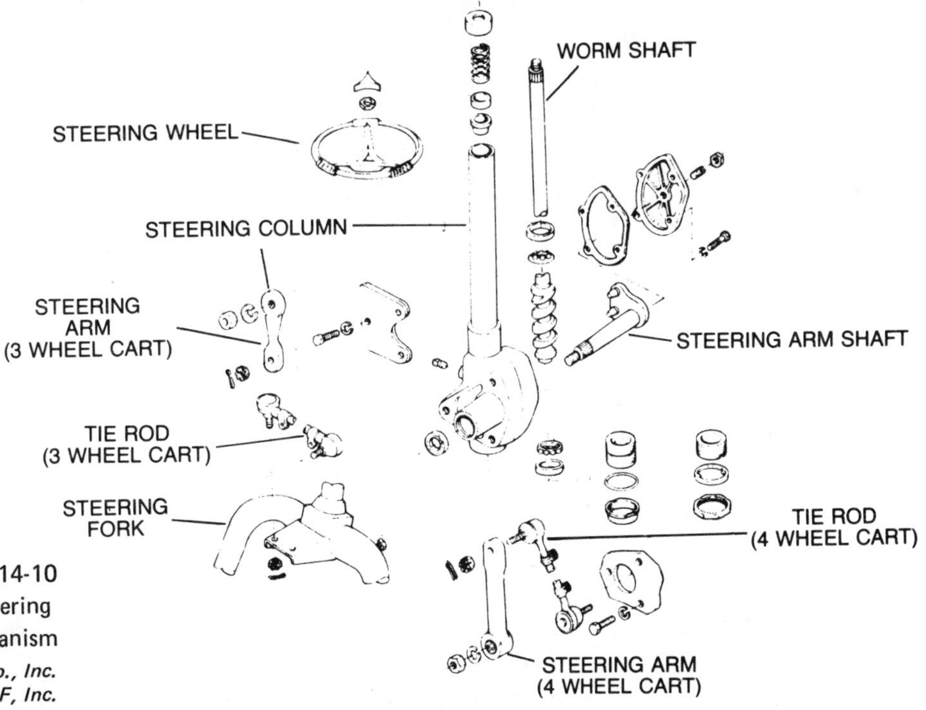

FIGURE 14-10
Exploded view of steering gear mechanism
Courtesy Harley-Davidson Motor Co., Inc. Subsidiary of AMF, Inc.

Most tractors use an I-beam type of front axle, which is supported by a pivot shaft in the center to the tractor front frame cross member. This arrangement permits each end of the axle to move up and down when going over rough ground. Springs are not used between the frame and axle. A spindle, which is part of the steering knuckle, supports the wheel. Steering knuckle arms are part of the assembly. King pins, sometimes called "steering knuckle bolts," attach and support the steering knuckle assembly to the axle and act as hinge pins so the knuckle assembly can swing back and forth.

A tie rod with a ball joint on either end connects the two steering knuckle arms (Figure 14-4). The tie rod and tie rod ends are threaded, so the end can be replaced, as well as shortened or lengthened, to adjust the toe-in of the front wheels. A drag link with a ball joint is used to connect the left steering knuckle assembly to the steering gear. Figure 14-11 shows a more complex front end assembly, in which each wheel is independently suspended. All steering systems with four wheels have steering knuckles, steering knuckle arms, a drag link, and tie rods.

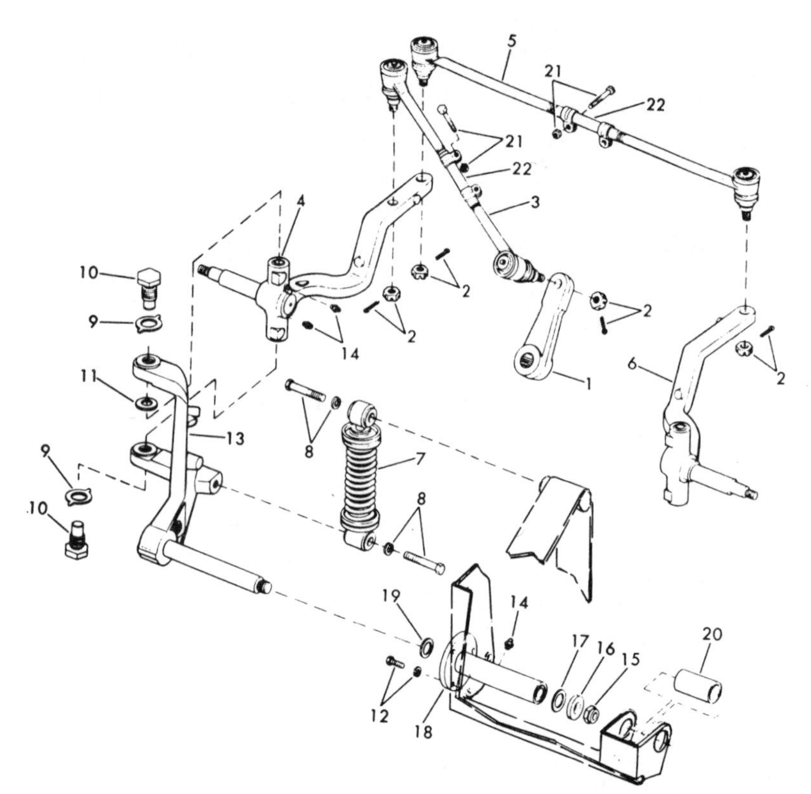

FIGURE 14-11
Exploded view of front suspension and steering
Courtesy Harley-Davidson Motor Co., Inc. Subsidiary of AMF, Inc.

1. STEERING ARM
2. CASTLE NUT AND COTTER PIN (4)
3. SHORT TIE ROD ASSEMBLY
4. RIGHT AXLE SUPPORT ARM
5. LONG TIE ROD ASSEMBLY
6. LEFT AXLE SUPPORT ARM
7. SHOCK ABSORBER (2)
8. BOLT AND LOCKWASHER (4)
9. TAB LOCKWASHER (4)
11. THRUST WASHER (2)
12. BOLT AND LOCKWASHER (6)
13. SUSPENSION ARM (2) (RIGHT AND LEF
14. GREASE FITTING (6)
15. NUT
16. SPACER
17. SHIM (2) (AS REQUIRED)
18. SUSPENSION TUBE (2)
19. SPACER (EARLY 1972 MODELS)
20. RUBBER SLEEVE
21. CLAMPING BOLT AND NUT (4)
22. ADJUSTING SLEEVE (2)

A much simpler steering gear assembly may be found on some tractors. A spur gear, mounted on the lower end of the steering wheel shaft, is in mesh with teeth on a drag link plate, which is mounted on the side of the frame. The drag link is attached to the drag link gear. As the steering wheel is turned, the drag link is caused to move back and forth. Figure 14-12 shows a simple steering linkage arrangement on a tractor.

FIGURE 14-12
Tractor steering linkage
Courtesy Engineering Products Co., Inc.

SNOWMOBILE STEERING

A snowmobile uses a handlebar to turn the skis. Each ski is supported on the chassis by a spindle, which turns in bushings. A steering arm is attached to the top of each spindle (Figure 14-13). A tie rod on each side connects the spindle arm to a steering arm on the lower end of the steering post. Figure 14-13 shows the steering spindles attached to the skis and frame of the body. The upper end of the steering post is attached to the handlebar. The tie rod must be adjusted to equal lengths on each side for the skis to track properly. The skis must be the same distance apart at the front as they are at the rear.

FIGURE 14-13
Snowmobile steering spindles
Courtesy Polaris E-Z-Go Division of Textron Inc.

WHEELS

Wheels are used to provide a rolling surface for a number of units. When the machine is self-propelled, they not only carry the machine's weight, but they are also used to propell it. When used to carry any amount of weight, wheels are mounted on some type of antifriction bearing of either the ball or roller type. When the load is not too great, a bronze bushing may provide a friction surface. Front wheel bearings, because the load angle varies as the wheels are turned for steering, must be properly adjusted for wear and kept lubricated.

Most wheels are steel stampings and designed for pneumatic tires. The center of the wheel is lower than the outside rim so the bead of the tire can be dropped into the center when removing the tire from the wheel. A hub is part of the wheel and is designed to contain bearings. Mopeds generally use wire spoke wheels, the same as on bicycles. The spokes can be adjusted to keep the rim and wheel assembly in alignment.

MOPED WHEELS

Both front and rear wheels of a moped are mounted on a spindle using ball or roller bearings to provide a free rolling surface. The bearings are adjustable by turning the cone, if threaded, or the cone lock nut. With a threaded cone, the lock nut must be loosened before turning the cone. The spindle should be centered in the hub. The bearings are adjusted so the wheel turns freely without any play. Figure 14-14 shows a moped spindle, hub, and bearing cone. All

FIGURE 14-14
Moped wheel spindle, hub, and bearing cone
Courtesy Motobecane America, Ltd.

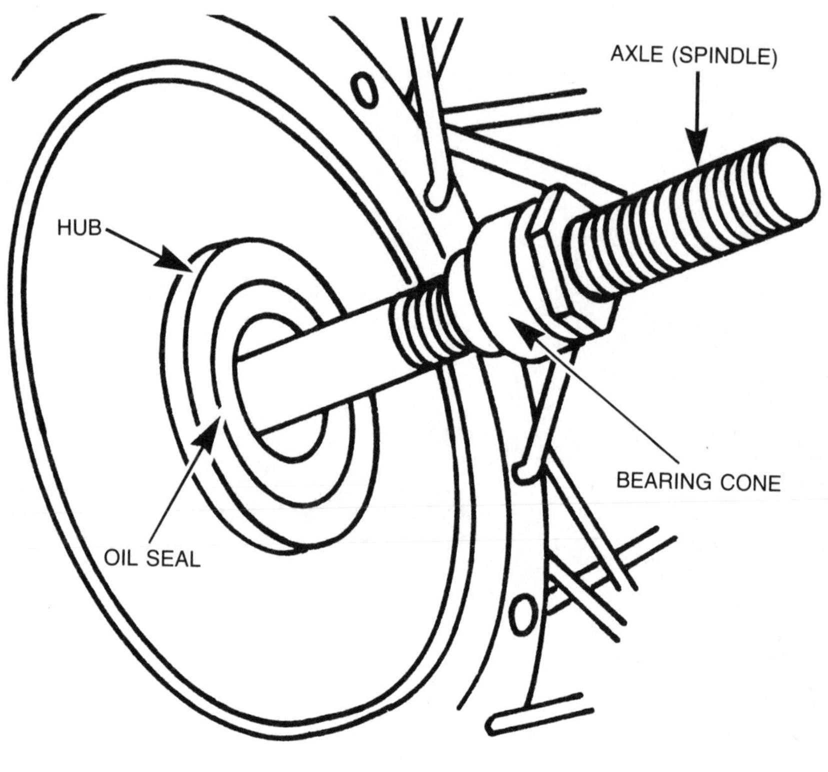

bearings must roll freely, and the balls or rollers must not be rough or chipped. The bearing cup and cone must be smooth and show no signs of wear. The bearings must be kept lubricated to give satisfactory service.

When being installed moped wheels must be properly located so as to track properly. The rear wheel is aligned by the chain-adjusting screws.

TRACTOR WHEELS

The rear wheels are generally mounted on a driving axle, which is supported on a bearing or bushing. Generally no adjustment is needed. The rear wheel side play (in and out) is usually adjusted by adding or removing washers of different thicknesses. Side play should normally not exceed 1/8 inch.

The front wheels of the heavier tractors are usually mounted on roller bearings. Light-weight tractors and riding mowers may use bushings. When the wheel is held onto the spindle with a nut, the bearings are adjustable by tightening or loosening the adjusting nut. The wheel should turn freely with no play. Figure 14-15 shows the front hub assembly using adjustable roller bearings, an installation that is used on a golf cart. When either a cap screw and washer or a cotter pin is used to hold the wheel to the spindle, excess play—more than 1/8 inch in or out movement—is removed by the addition of washers.

If you need to remove the hub for any reason, inspect the bearings, if they are the ball or roller type, for chips, wear, or roughness and replace them if necessary. The bearing cup is pressed into the hub. The bearing cone or roller is a slip fit on the spindle. Always replace the grease seal when replacing the bearings. Replace the entire bearing assembly even if only one component is faulty. Always repack the bearings with grease when reinstalling them.

The idler wheels used with a snowmobile track are mounted and turn on shafts. They need to be kept properly lubricated. Should excess free play develop, the bearings should be replaced.

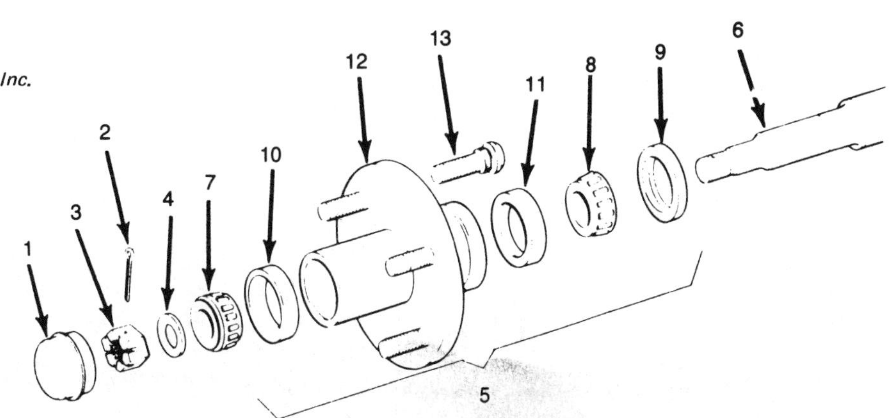

FIGURE 14-15
Front wheel hub, bearings, and spindle

Courtesy Harley-Davidson Motor Co., Inc. Subsidiary of AMF, Inc.

1. GREASE CAP
2. COTTER PIN
3. AXLE NUT
4. WASHER
5. WHEEL HUB ASSEMBLY
6. AXLE-SPINDLE
7. OUTER CONE BEARING
8. INNER CONE BEARING
9. OIL SEAL
10. OUTER BEARING CUP
11. INNER BEARING CUP
12. WHEEL HUB
13. WHEEL STUD

TIRES

The tires act to cushion the ride and provide driving traction. For good tire wear, the tires must be kept properly inflated. The air pressure required is generally marked on the tire sidewall. For longer tire life, it is better to over-inflate than to operate with too little air pressure.

Also inspect the tires regularly to see if they have picked up a nail or other object that may work itself into the tire and cause a leak. If there is an irregular wear pattern, the suspension and steering system should be inspected for malfunctions and misalignment.

CHECKING SUSPENSION SYSTEMS

A weak or broken spring is easy to locate, because the machine tends to lean toward the weak side. A broken leaf or coil can readily be seen.

Shock absorbers are designed to counter spring rebound. To check the shock absorber action, bounce the machine up and down by hand. After releasing the machine it should immediately come to rest; if not, the shocks are faulty. Generally leakage causes a shock to fail. So check the shock for leakage around the piston rod. If oil is present, the shock is leaking. When replacing a shock absorber, always install new rubber bushings.

Check the entire suspens system for loose bolts and any signs of misalignment or wear.

The ski limiter and the rear spring tension can be adjusted on most snowmobiles to compensate for load and terrain conditions. Usually two or three bolt holes are located at the points where the rear suspension is attached to the machine (Figure 14-16). Changing the location of the suspension arms in the mounting holes has a bearing on the riding and handling characteristics.

The pivot arm for the front ski usually has two or more holes so the angle of the ski can be changed (Figure 14-17). The center setting is ordinarily the proper setting for most operational conditions.

Track slack should be kept to a minimum. "Minimum" is generally less than 1/4 inch clearance, measured between the slide rail

FIGURE 14-16
Snowmobile spring adjustment locations
Courtesy Polaris E-Z-Go Division of Textron, Inc.

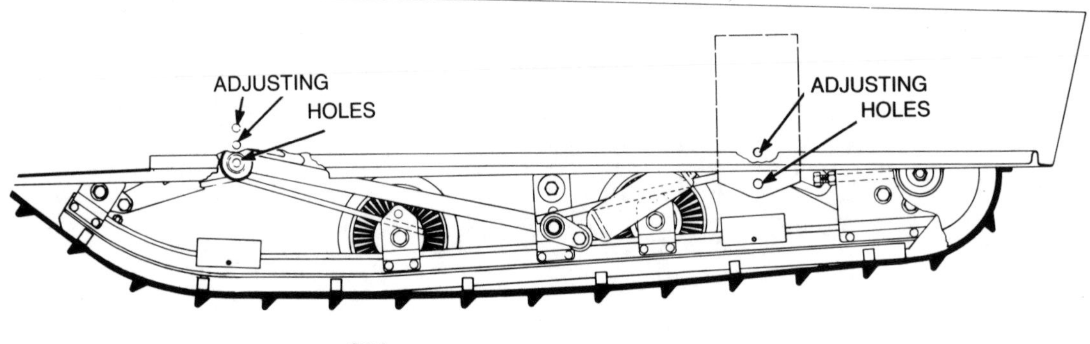

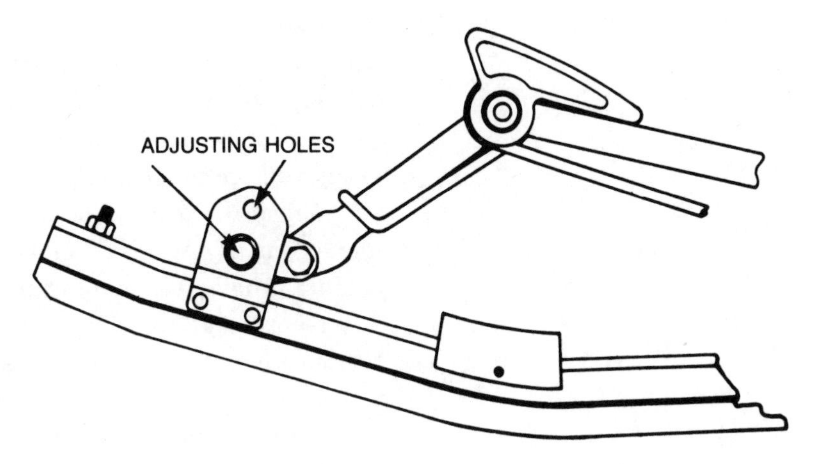

FIGURE 14-17
Snowmobile front ski adjustment
Courtesy Polaris E-Z-Go Division of Textron, Inc.

and track clip when the rear end of the machine is off the ground (Figure 13-12). The idler shaft can be adjusted to compensate for wear. Check the manufacturer's specifications for the exact settings.

The entire rear suspension system must be kept lubricated to prevent wear. Replaceable wear bars are used on the track, as well as on the skis.

CHECKING STEERING SYSTEMS

With the front wheels in a straight-ahead position, move the steering wheel back and forth a small amount. This test determines if there is free play due to wear or adjustment in the steering gear mechanism and linkage. By watching the linkage while moving the steering wheel, you can determine where the wear is located.

If the lost motion is in the steering gear assembly, look for an adjusting screw on the gear box. After loosening the lock nut, turn the adjustment to remove the free play.

If the ball and socket joints are loose on the tie rods or drag link, the joint can usually be removed and new ones installed. The tie rods are adjustable and must be turned in or out of the tie rod ends to obtain the correct toe-in if the front wheels do not line up properly. The correct toe-in is usually 1/8 inch. In other words, on most installations, the fronts of the tires are 1/8 inch closer together than the rears. Take the measurement using a straight edge about one foot off the ground both at the front of the inside of the tire and at the rear inside of the tire. The straight edge must be kept level. One tie rod generally has left-hand threads, while the opposite end has right-hand threads—so the rod can be shortened or lengthened by turning.

CHECKING WHEEL ALIGNMENT

Raise the front end of the machine off the ground, and spin each wheel. They should turn freely without making any noise. If there is any roughness, noise, or binding, then remove the wheel and inspect the bearings or bushings. If antifriction bearings are used, check the balls or rollers, as well as the cup and cone, for any signs of wear, chipping, or roughness. Replace the entire bearing assembly if any wear or chipping is indicated. If bushings are used and free play exists, replace the bushings.

Grasp the top and bottom of the tire and try to move the wheel in and out. If the wheel moves in and out on the spindle, adjust the bearing. The wheel should turn without binding, and there should be no free play. If an adjusting nut holds the wheel on the spindle, tighten the nut to remove the excessive play. If a cotter pin holds the wheel to the spindle, or if a cap screw and washer are used, remove the cotter pin or cap screw and install washers to remove the excessive free play.

Grasping the top and bottom of the tire, pull out on the top of the tire and push in on the bottom. Free play here means the bolt and bushing, or the pin and bushing, that hold the steering knuckle assembly to the axle are worn. New pins and bushings need to be installed. After installing new bushings, you generally need to ream or hone them. An automotive machine shop can do so for you.

TIRE REMOVAL

If a tire goes flat, remove it from the machine by removing the mounting nuts. On a moped, the brake control cables and chains must be removed before removing the spindle nuts, which hold the wheel spindle to the fork. After removal inflate the tire to approximately 20 psi. Immerse the tire in water to determine where it leaks, as indicated by bubbles. Mark where the bubbles escape. The leak could be due to a puncture, a leaking valve core, or a valve stem that is not properly sealing at the rim.

If puncture is small, put in a plug from the outside to seal the leak without removing the tire from the wheel. Any service station has the equipment to do this job.

If you have to remove the tire from the wheel, proceed in the following manner. Remove the valve core. Loosen both tire beads from the rim flanges by stepping on the tire sidewalls. A rubber mallet might be of assistance while stepping on the tire. Using a tire tool, carefully pry the upper bead over the edge of the wheel rim. Pry a small amount at a time. Do not use excessive force, or you might damage or stretch the bead. If the bead is damaged it will not hold air. When one bead is removed, make sure the remaining bead is in the rim well (center). When the bead is started over the rim flange it should pry off with no trouble (Figure 14-18).

When either the problem has been taken care of or another tire is to be installed, clean both tire beads to remove all traces of dirt and foreign matter. Clean the rim flange as well. Apply a liberal amount of tire mounting solution to the tire bead and rim. Install the tire on the wheel using a rubber mallet and tire iron. Shake the tire to get beads out of the center of the wheel and out against the edge of the rim flange. Set the tire against the wall in an upright position. Apply high-pressure air through the valve while pushing against the tire. This should help the air pressure to force the beads out against the wheel rim. When the beads snap into place, remove the pressure hose and install the valve core. Correct the tire pressure and immerse the tire in water to check for leaks. Then reinstall the wheel.

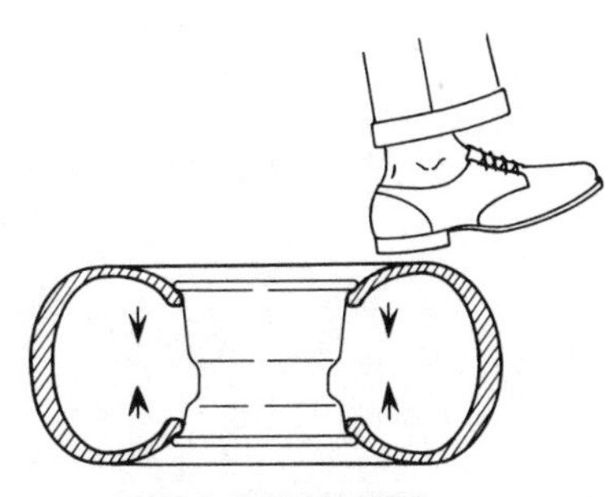

STEP I. BREAK TIRE BEADS FREE OR RIM

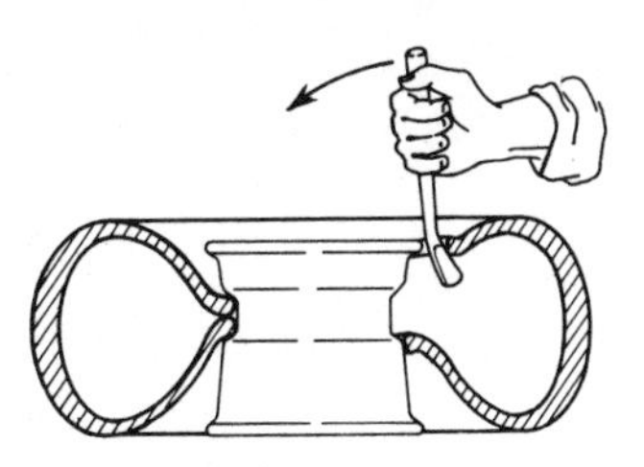

STEP II. REMOVING UPPER BEAD FROM RIM

STEP III. REMOVING LOWER TIRE BEAD FROM RIM

INFLATING TIRE

FIGURE 14-18
Tire replacement
Courtesy Harley-Davidson Motor Co., Inc. Subsidiary of AMF, Inc.

15 BRAKE SYSTEMS

DISC BRAKES
DRUM BRAKES
BAND BRAKES
BRAKE LINKAGES
HYDRAULIC BRAKES
CHAIN SAW BRAKE
BRAKE SERVICE

A brake system is necessary to stop the machine, assist in controlling the speed of travel, or to hold the machine stationary. In a chain saw a brake is generally used to stop the chain from turning when kickback occurs.

Since any mass in motion represents energy, to stop the motion, this energy must be absorbed or converted to some other form. A brake system does so by converting the energy of motion into heat.

On a more practical level, brakes may be considered mechanical devices that use the force of friction to overcome motion. Since any two surfaces in contact with one another resist movement between them, the moving brake drum or disc is slowed by applying pressure from a non-rotating surface. The nonrotating surface is generally made of an antifriction material composed primarily of asbestos.

In tractors, riding mowers, mopeds, golf carts, snowmobiles, and other self-propelled machines that carry people, the brake system must be capable of slowing down the machine, stopping it, and then holding it in a stationary position. While automobiles must have two separate brake systems, a service brake and a parking brake, off-the-highway machines, including mopeds, only use one. The brake that is used to stop the machine is also used to hold the machine in a stationary position. The hydraulic brake system is an exception to this. A few machines use a hydraulic pump (master cylinder) to force fluid to a hydraulic cylinder. Pistons in the cylinder expand the brake shoes, bringing the lining into contact with the drum.

While hydraulic pressure applies the brakes, the parking brake still uses cables and/or rods to keep the brake applied. The service brake is usually applied by a foot-operated pedal. The pedal actuates rods, or in some cases rods and cables, to bring the brake lining into contact with the revolving drum or disc. On some machines the same foot-operated pedal is used to operate both the clutch and the brake. Pushing the pedal part of the way down disengages the clutch; pushing it beyond this point applies the brake. A hand-operated lever, or on some installations a ratchet on the foot pedal, is used to hold the brake in an applied position.

Most brake systems are simple and easy to understand, as well as to adjust and repair. The following information is general and applies to most brake systems. Three types of brake arrangements are in general use:

1. The most common type is the *disc brake.* Either a disc is located at each rear wheel, or just one disc-type brake is used at the transmission or drive shaft. With a single disc, the braking effort is multiplied due to the gear reduction in the rear axle assembly.

2. A *drum brake* assembly is used on some machines. The most common type uses shoes with a lining attached and located inside the drum. When the brakes are applied, the shoes are forced outward against the drum.

3. A third type brake assembly uses a *flexible band,* with a lining riveted to its inside, that clamps around the outside of a drum.

DISC BRAKES

On a number of tractors the disc brake is located at each rear driving wheel, although in some cases it is located inside the transmission case. Other disc brake assemblies may be located on the drive shaft, ahead of the differential assembly. In this last case, the braking effort is multiplied by the gear ratio of the ring gear and drive pinion. Regardless of its location, all disc brakes operate in pretty much the same way.

The disc brake in common use is the caliper type; the brake shoes, when applied, clamp onto both sides of the disc. A brake shoe, sometimes called a "brake pad" or a "brake puck," is located on either side of the disc, mounted on anchor pins or bolts so it can be moved into contact with the disc. When the brake pedal is depressed, it transmits a pulling force to the caliper, which creates a clamping force on the disc through a cam and lever arrangement. A spring arrangement holds the lining away from the disc when the brakes are released.

Figure 15-1 shows a disassembled, as well as an assembled, view of a caliper-type disc brake used on a golf cart. This brake is cable-operated and can be adjusted by shortening or lengthening the cable.

FIGURE 15-1
Caliper-type disc brake — assembled and disassembled
Courtesy Harley-Davidson Motor Co., Inc. Subsidiary of AMF, Inc.

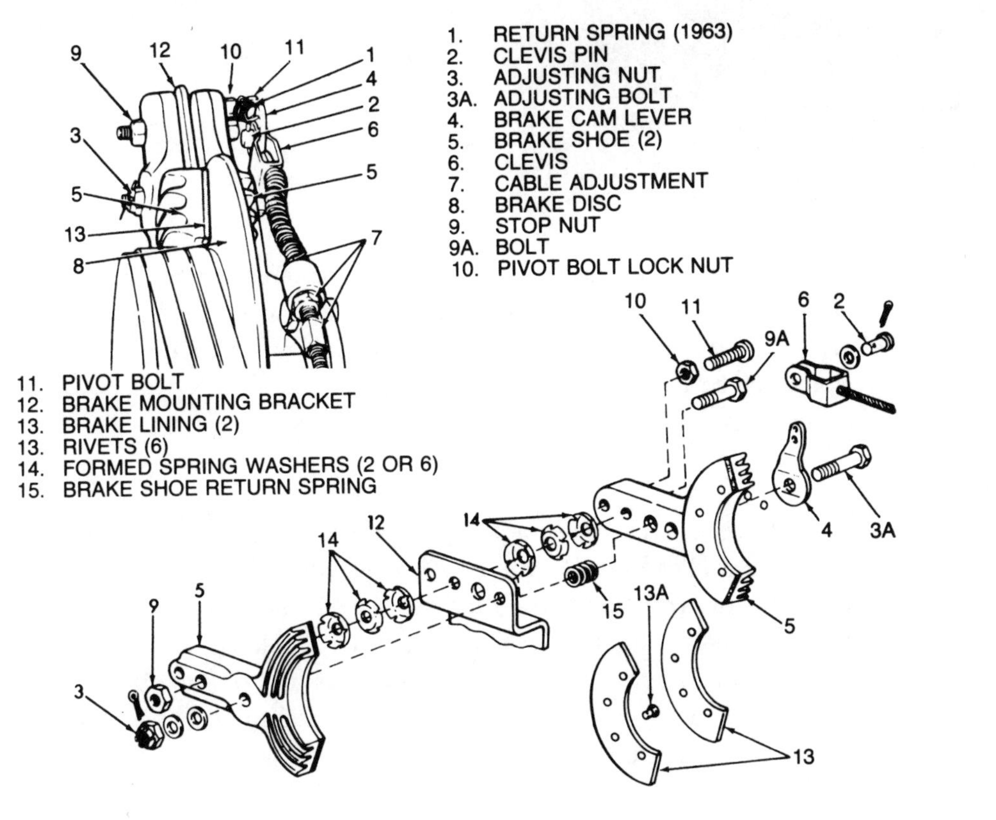

The caliper-type brake itself is not adjustable in most cases, but the operating rod or cable can be lengthened or shortened to compensate for wear. The lining can be replaced when it becomes worn to the extent that the lining rivet heads are within 1/32 inch of the lining face.

Some disc brakes are not of the caliper type. The brake installed inside the transmission case or in the differential housing ahead of the drive pinion may use a disc that can slide on the shaft on which it is mounted. In this type of construction, which may also be used on the driving shaft of some snowmobiles, pressure is applied to one shoe and lining. This pressure forces the disc to move into contact with a stationary lining on the other side of the disc. Figure 15-2 shows a disc brake used on some snowmobiles. The brake is applied by a lever on the handlebar, which is connected by a cable to the cam lever on the brake-actuating mechanism.

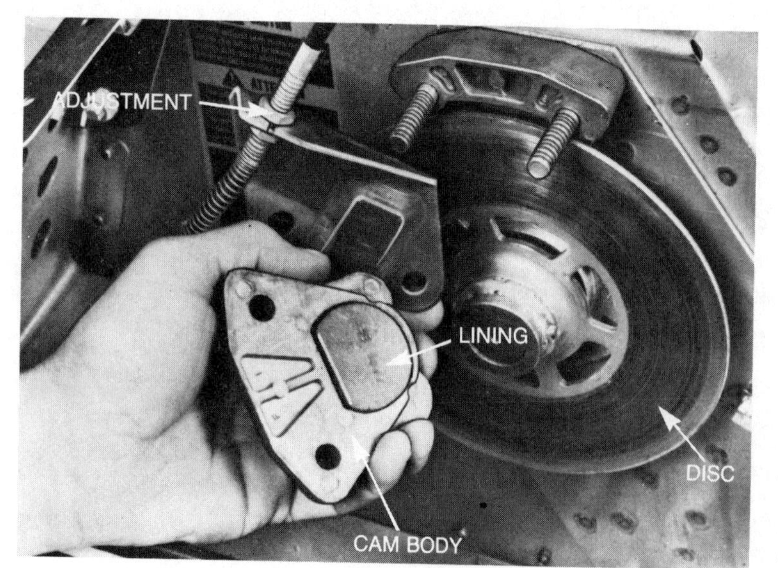

FIGURE 15-2
Disc brake cam body, lining, and disc
Courtesy Deere and Company, Moline, Illinois

The brake is adjusted by shortening or lengthening the cable at the disc. Some brakes of this type may have an adjusting screw on the cam to move the cam shoe and lining closer to the disc. The lining can be replaced on this type of brake.

DRUM BRAKES

A number of machines are equipped with a drum-type brake—mopeds, golf carts, and various other machines. A drum is attached to and turns with the wheel, and two brake shoes with their linings are mounted on the brake backing plate. The backing plate is fastened solidly on a frame member or housing. One end of each brake shoe is held against a solid anchor by springs, and the other end of the shoes are held against a cam by spring tension. A lever is attached to the cam.

Figure 15-3 shows a brake assembly used on a moped. In a moped, a cable and/or rods connect the lever to the brake pedal or lever. When the pedal or lever is depressed, the lever rotates the cam. The cam forces the brake shoes apart, bringing the lining into contact with the revolving drum. As more pressure is exerted on the pedal, the force exerted on the drum becomes greater, and the stop is quicker.

Shortening or lengthening the rod or cable brings the shoes and lining into the proper adjustment. The shoes should be adjusted so there is just enough clearance between the drum and lining to prevent drag. The lining can be removed from the shoes and replaced, if it becomes so worn that the rivet heads are about to contact the drum. It is generally accepted practice to exchange the old shoes for shoes with a new lining.

Replacing the shoes is generally a simple matter. Either unhook the springs or pry the shoes off the backing plate, and then remove

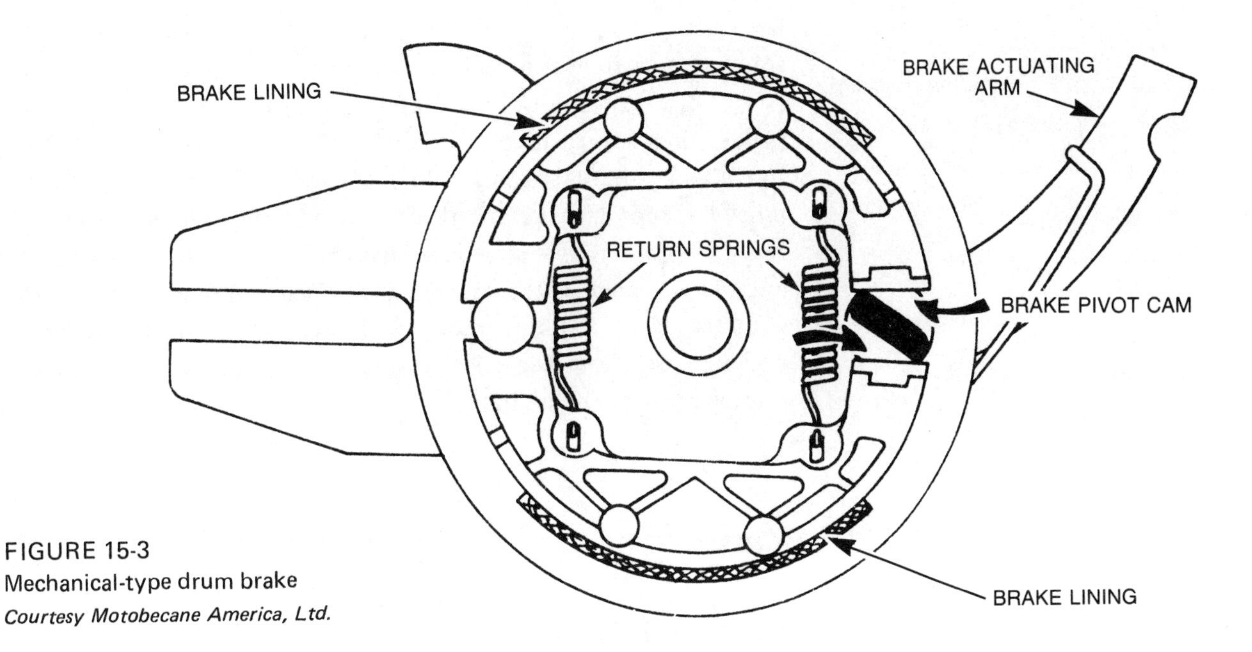

FIGURE 15-3
Mechanical-type drum brake
Courtesy Motobecane America, Ltd.

the springs. On some installations the tension of the springs may vary. Mark the springs so they can be put back in the same relative position; do the same to the brake shoes, because there may be a difference in the type or length of lining.

Some machines use a hydraulic brake system. The difference in the brake mechanism is that a hydraulic cylinder is used between the ends of the brake shoes in place of a cam. Figure 15-4 shows a disassembled view of a hydraulic brake. An adjuster, between the ends of the shoes opposite the cylinder, is threaded so it can be turned in or out to move the shoes closer to or farther away from the drum. A

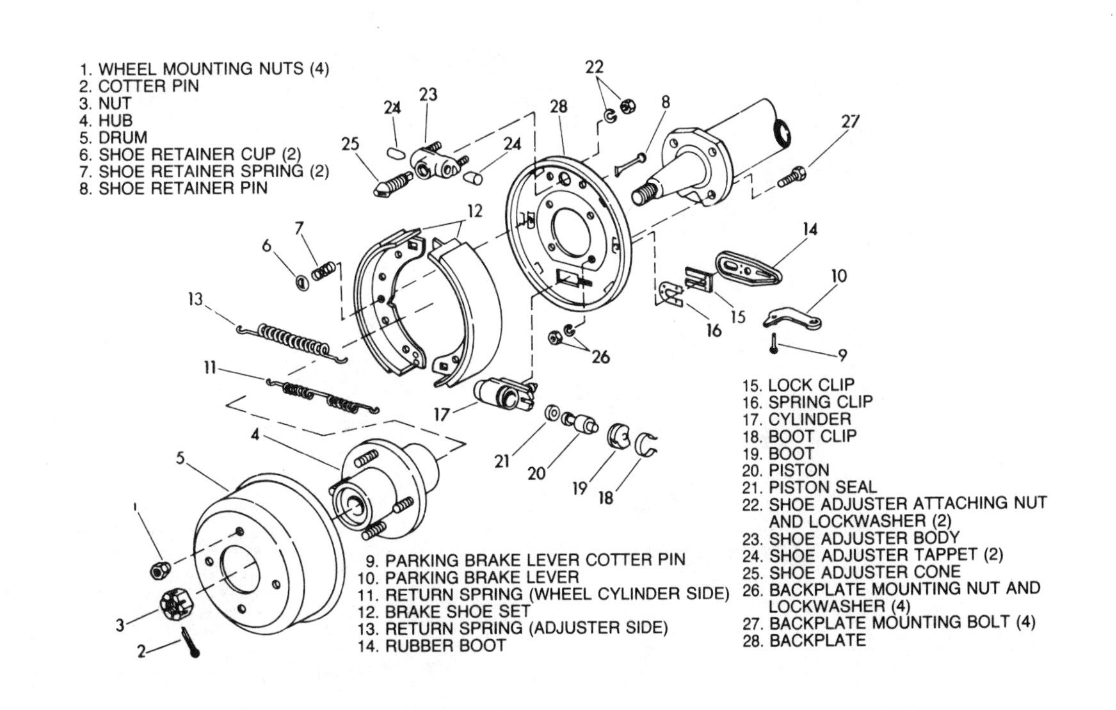

FIGURE 15-4
Disassembled hydraulic brake unit
Courtesy Harley-Davidson Motor Co., Inc. Subsidiary of AMF, Inc.

lever, actuated by the parking brake lever, and a cable forces the shoes against the drum to hold the machine in park.

When working on the brake or wheel, do not step on the brake pedal with the drum off, because the hydraulic pressure will force the wheel cylinder piston out of the cylinder. The hydraulic system must be kept free of air. If air gets into the system, the pedal gets spongy. To remove air, open up the bleeder valve on the wheel cylinder, a small valve on the side of the cylinder. With the master cylinder full of fluid, pump the brake pedal until bubbles stop coming out. Close the bleeder valve when a stream of fluid is being pumped out of the cylinder.

BAND BRAKES

Some machines use a flexible band type brake on the rear wheels, on the drive shaft, or on the transmission shaft. When used on the transmission or drive shaft, one band is generally all that is needed. The brake drum is attached to and turns with the rear wheels, transmission shaft, or drive shaft. A flexible steel band, with a flexible brake lining riveted onto it, fits around the outisde of the drum. The ends of the band are fastened through a linkage to a bracket, which is attached solidly to a frame member. A spring arrangement holds the band off the drum. A lever arrangement is part of the linkage. When the brake pedal is depressed, a set of rods or a cable actuates the lever on the linkage, which clamps the band tightly around the revolving drum. Friction stops the drum when enough pressure is applied. Figure 15-5 shows this type of brake as used on a tractor.

FIGURE 15-5
Flexible band brake used on a tractor
Courtesy Engineering Products Co., Inc.

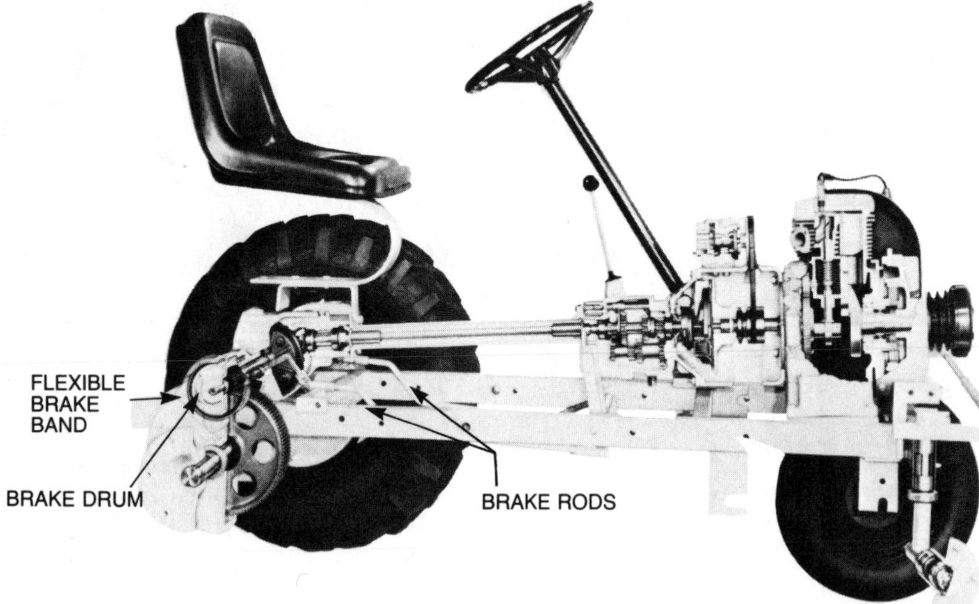

Some chain saws use a flexible brake band around the clutch drum. When the anti-kickback handle is knocked back, the flexible band applies pressure around the drum, stopping the chain. A lining may or may not be used on this band (Figure 15-10).

BRAKE LINKAGES

To apply the brake, the movement of a pedal or lever must be transmitted to the brake shoe by means of rods, cables, levers, and, in the case of some machines, hydraulic pressure. For ease of operation, the system must contain a mechanical advantage, which is usually brought about by leverage. The brake pedal is a lever with about a 6-to-1 mechanical advantage. The lever, which actuates the cam or linkage, provides further mechanical advantage.

Figure 15-6 shows a typical brake pedal using a cable arrangement for applying the brakes. Figure 15-7 shows a rod-and-lever arrangement. In both cases the rods and cables may be shortened or lengthened. A ratchet arrangement is used on some pedals to hold the pedal in an applied position in park.

Many mopeds and snowmobiles have a lever by the grip on the steering handlebar to apply the brake. The lever actuates a cable to apply the brake mechanism, whether it is of the mechanical or hydraulic type. Figure 15-8 shows a brake pedal used on a moped.

Some tractors use one pedal to control both the clutch and the brake. The clutch is released during the first part of pedal travel; then after the clutch is released, the brakes are applied. In this simple arrangement the brake linkage is not actuated until the pedal is par-

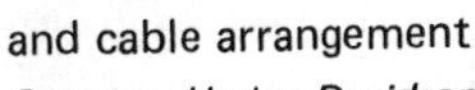

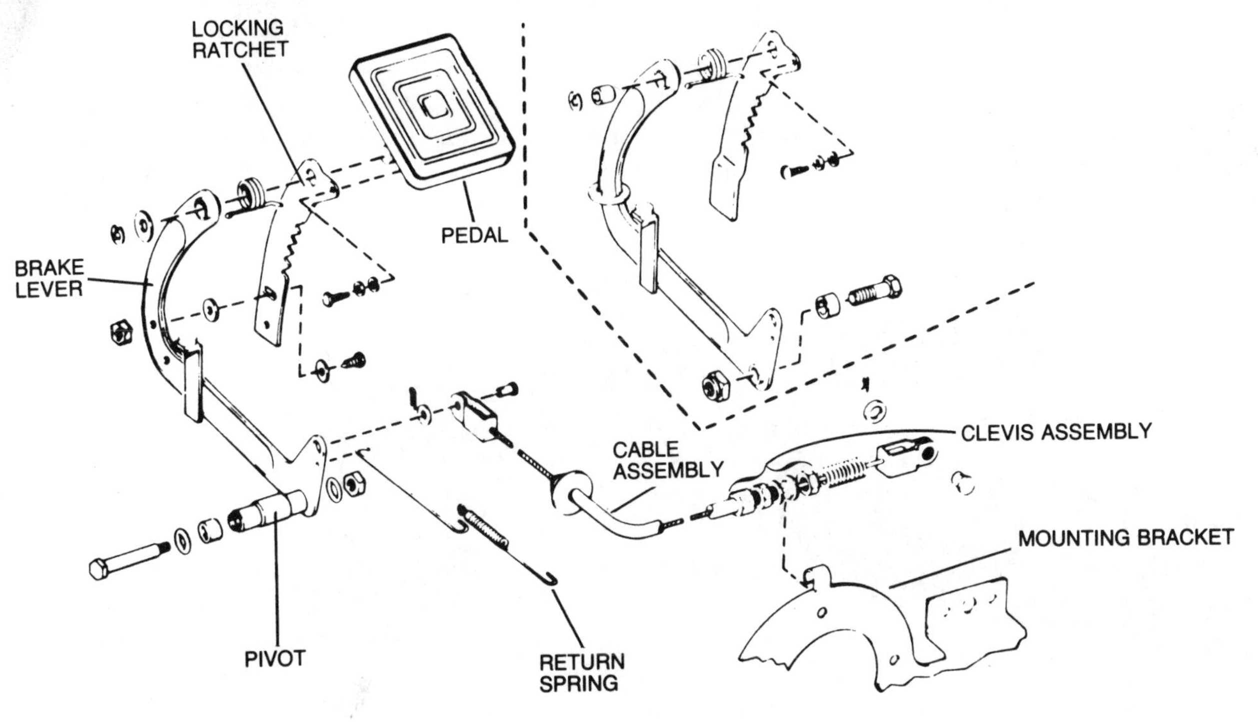

FIGURE 15-6
Exploded view of brake pedal and cable arrangement
Courtesy Harley-Davidson Motor Co., Inc. Subsidiary of AMF, Inc.

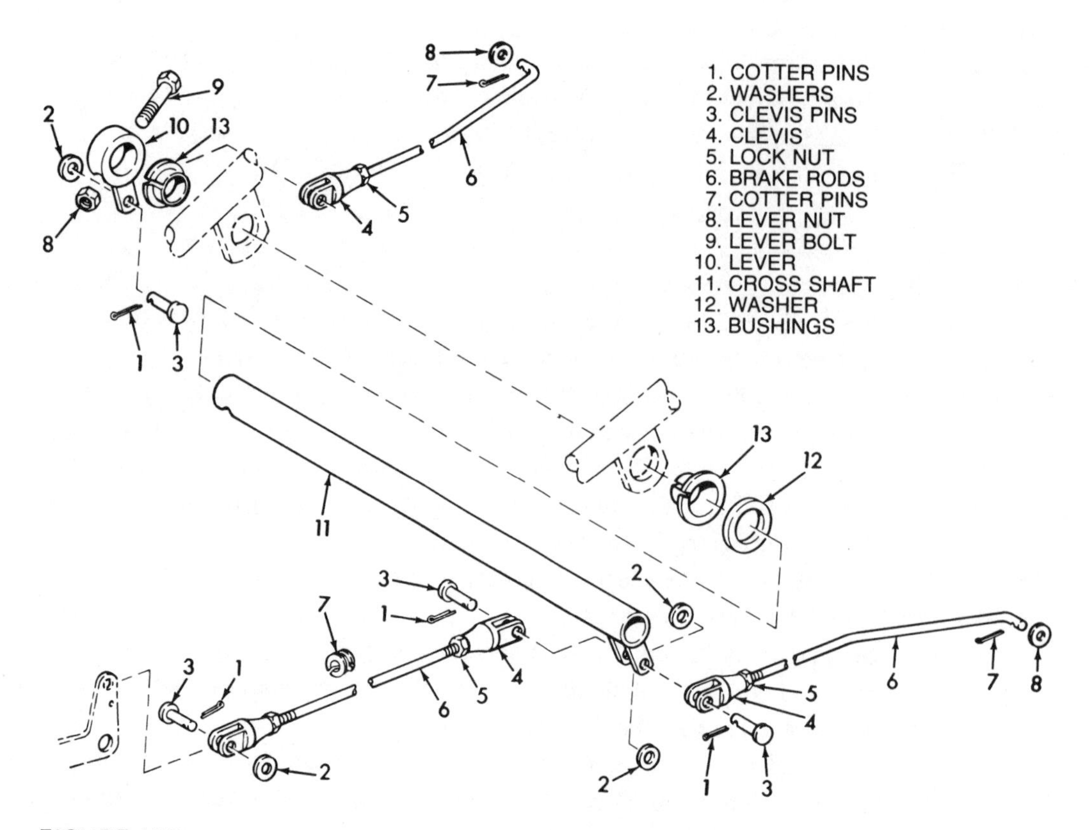

FIGURE 15-7
Mechanical brake linkage
Courtesy Harley-Davidson Motor Co., Inc. Subsidiary of AMF, Inc.

tially depressed. The two actions are compatible since the clutch should be released when stopping anyway. (Before shifting gears on a tractor, you should come nearly to a stop.) With practice you soon "feel" when the clutch is released and when the brakes begin to apply. The linkage adjustment is important on this type of installation.

In many cases the parking brake is applied by a ratchet on the foot pedal. Pushing down on the top edge of the pedal moves a lever with notches on the edge into engagement with a metal strip just below the pedal travel. This holds the pedal depressed. Pushing on the bottom part of the pedal releases the ratchet (Figure 15-6).

FIGURE 15-8
Snowmobile brake lever
Courtesy Polaris E-Z-Go Division of Textron, Inc.

Other machines may have a hand-operated lever that applies the brake when pulled back. A small ratchet holds the lever in the applied position. A plunger on top of the lever is generally used to release the brake. Pushing down on the plunger disengages the ratchet, letting the lever move forward and thus releasing the brakes. Applying the foot brake usually makes it easier to apply the hand brake.

HYDRAULIC BRAKES

A few machines use a hydraulic brake system. The hydraulic brake uses the same type of brake shoe, disc, or drum as the mechanically operated brake. But the brake-actuating mechanism consists not of cables and/or rods and a cam, but rather of hydraulic pressure. This pressure is created by a hydraulic pump (master cylinder), which delivers fluid under pressure to a cylinder located at the brake shoes or pad mechanism. To stop the machine, a piston or a number of pistons in the cylinder forces the brake shoes or pads into engagement with the drum or disc.

Figure 15-9 is an exploded view of a master cylinder. A piston, actuated by the brake pedal, forces hydraulic fluid under pressure into the brake lines to the wheel cylinders. An exploded view of the wheel cylinder is shown in Figure 15-4. Hydraulic fluid pressure on the piston in the cylinder forces the piston against the end of the brake shoe, thus creating friction between the shoe and drum to stop the machine. The same type of system operates equally as well with either the drum or disc brake.

FIGURE 15-9
Exploded view of master cylinder
Courtesy Harley-Davidson Motor Co., Inc. Subsidiary of AMF, Inc.

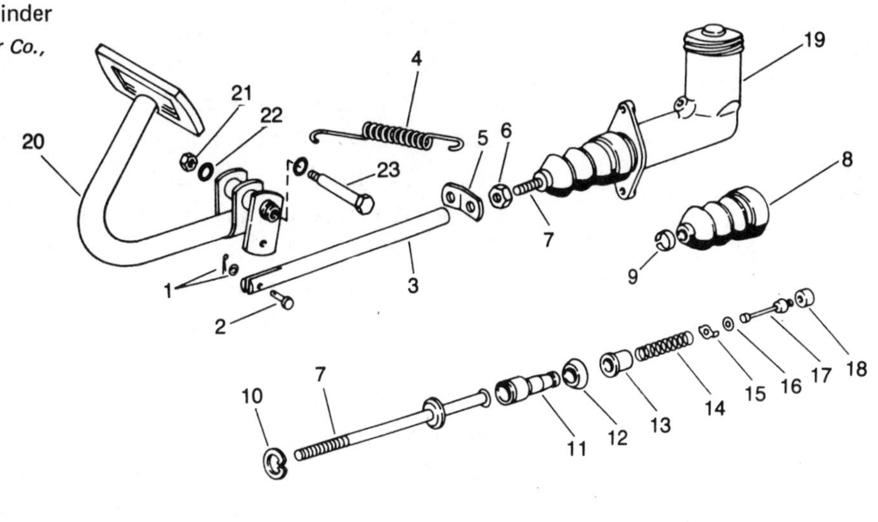

1. COTTER PIN AND WASHER
2. CLEVIS PIN
3. CLEVIS ROD
4. RETURN SPRING
5. RETURN SPRING BRACKET
6. CLEVIS ROD LOCK NUT
7. PUSH ROD
8. BOOT
9. BOOT RETAINER CLIP
10. PUSH ROD RETAINER CLIP
11. PISTON
12. PISTON SEAL
13. VALVE SPRING COLLAR
14. VALVE SPRING
15. VALVE SPACER
16. VALVE WASHER
17. VALVE STEM
18. VALVE SEAL
19. MASTER CYLINDER
20. BRAKE PEDAL
21. PIVOT BOLT NUT
22. PIVOT BOLT WASHER (2)
23. PIVOT BOLT

The important characteristic of a hydraulic brake is that the entire system must be kept air- and fluid-tight. If air gets into the system, the pedal reaction becomes spongy, because air compresses under pressure. Any leakage causes brake failure, because the pressure, rather than operating the cylinder, forces the brake fluid out of the system. A special hydraulic fluid must be used in the system. (Any petroleum-base fluid destroys the rubber parts.) Any time the system is opened, air can get into the system, and it must be bled out before the brakes can operate properly. If the fluid level in the master cylinder is not maintained, air gets into the system. But the cap can be removed from the master cylinder fluid reservoir without letting air into the system.

CHAIN SAW BRAKES

Most chain saws use anti-kickback devices, which commonly takes the form of a brake that operates to stop the chain when kickback occurs. Kickback happens when the chain hangs up, generally at the outer end of the bar. When the chain hits a solid object, or when it takes too large a cut, the chain catches in the wood and locks for an instant. The engine torque is then transferred to the guide bar, and the saw tries to turn upward and toward the operator.

To prevent kickback, some saws have a one-piece removable guard, similar to a large thick washer, that is wider than the chain. This is fastened to the nose of the bar to prevent the chain from hitting any thing as it passes the nose. The guard must be removed if you want to use the full length of the bar.

The most common anti-kickback device does not prevent kickback. But it does avoid injury to the operator by stopping either the chain, the engine, or both in a fraction of a second after the kickback occurs. The mechanism is a flexible brake band that goes around the outside of the clutch drum. The band, when actuated by the movement of the hand guard, unlatches the brake causing the band to clamp around the clutch drum by the force of a preloaded spring. At the same time a release plate disengages the driving plate from the hub and interrupts the flow of power between the crankshaft and clutch. The chain is thus stopped. Figure 15-10 shows this type of brake band installation.

Some installations may use a brake shoe and a cam. The shoe comes into contact with the clutch drum when the hand guard is activated. The same effect is achieved as when a band is used.

The flexible band may or may not use a lining on the friction surface. The brake must be reset (released) after it has stopped the chain. If the clutch and brake housing have vent openings, the mechanism must be inspected regularly and the saw dust and dirt removed.

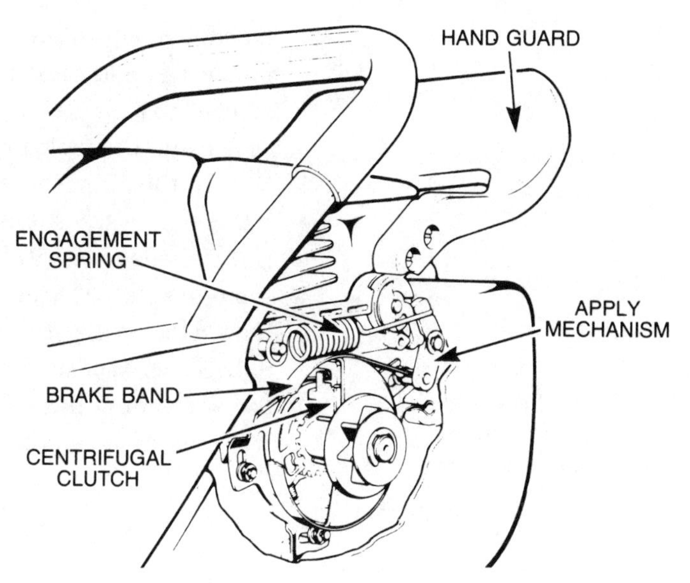

FIGURE 15-10
Chain saw brake
Courtesy Stihl Corporation

BRAKE SERVICE

Brake wear occurs with use, but the brake system is comparatively simple to service. The major point of wear is the lining, which wears every time the brakes are applied. So all brakes that are used to stop a propelled machine have lined brakes along with provisions for adjustment.

Some chain saw anti-kickback brakes are not adjustable. But if the saw has a band with a lining, the band can be replaced if the lining becomes worn. If a brake pad is used, the pad can be replaced.

When the brakes are properly adjusted, there should be approximately 1 to 1-1/2 inches of pedal travel remaining when the brakes are fully applied. In other words, the pedal should still have this much distance to travel before it contacts a stopping surface, such as a floorboard. If this much margin is left, if the wheels appear to stop evenly, and if there is no noise when the brakes are applied, the system is functioning properly. But if the pedal travels too far before the brakes take hold, if the brakes pull unevenly, or if there is noise when the brakes are applied, then the system requires service.

If a noise occurs when the brakes are applied, a rivet could be contacting the drum or the lining could be so worn that the metal shoe is in contact with the drum or disc. To check for even brake application when the brakes are used on both rear wheels, raise the wheels off the ground. With the engine in gear and operating at a fast idle, apply the brakes. If one wheel stops turning before the other, adjust the brake linkage on the "late" wheel until it stops at the same time as the other wheel.

Inspect the lining for wear. To do so on the drum-type brake you have to remove the drum. There should be approximately 1/32 inch of lining material left before the rivet heads can contact the drum or disc. At the same time the drum can be checked for grooves, roughness, and scores. If the drum is badly grooved and/or scored, it should be remachined on a brake drum lathe. The drum should also be checked for out-of-round, a test that requires a special brake tool. Drum and disc truing can be done at an automotive machine shop.

If the lining is worn, replace it. First disconnect the springs; then remove the shoes. A pliers will enable you to stretch the spring, and a screwdriver can be used to pry the spring off the retainer. On some brakes you can simply pry the shoes off the backing plate and then remove the springs. Mark the springs and shoes so they can be replaced in the same relative positions. Figure 15-11 shows removing the shoes on a moped brake assembly. The wheel must be removed from the frame and the backing plate removed from the hub.

FIGURE 15-11
Removing brake shoes
Courtesy Motobecane America, Ltd.

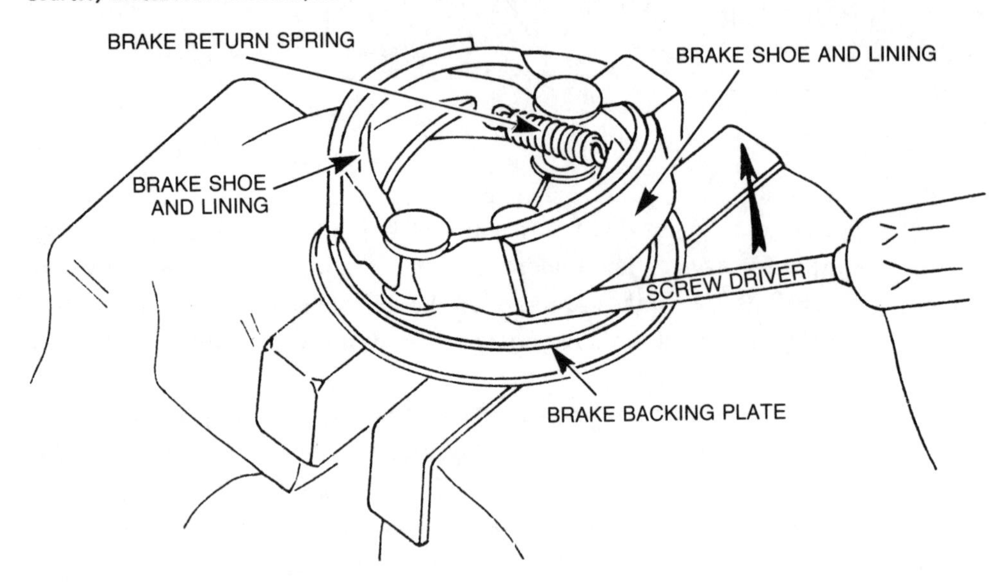

If the lining is worn so that it needs to be replaced, exchange the shoes and lining for relined shoes.

On the disc brake you have to remove the shoe, pad, or puck to determine the exact condition of the lining. Normally you can see the edge of the lining. If the lining is still 3/16 to 1/4 inch thick and if the disc is not badly scored, the lining should still be usable. The removal of the pad, shoe, or puck is generally apparent. Take the brake assembly apart and install new linings, which usually can be purchased on an exchange basis.

Figure 15-2 shows a disc brake assembly on a snowmobile. On this noncaliper brake, the disc slides on the shaft. The cam body is removed. The disc is pulled off the shaft after removing the retainer, and the inner puck can be removed. In some cases you may have to heat the puck to loosen it for removal. Some cam bodies may have an adjusting screw, which permits you to move the puck closer to the disc when excessive clearance exists.

Many brake assemblies are not adjustable, so the free play is reduced by shortening the cable or rods. In every case when adjustment is possible, the lining is moved as close as possible to the drum or disc without creating a drag.

When levers are used, the linkage should always be adjusted so that, when the brake is applied, the levers are not over center, that is, they should not be beyond perpendicular with the pivot point. If they are, they cause a hard brake action because they are pulling against themselves.

The adjusting screws for the hydraulic brake are shown in Figure 15-4. The screws can be turned from the back side of the backing plate.

A bleeder valve is located on the hydraulic cylinder, so air can be bled from the system should it become trapped someplace within the system. When bleeding brakes, be sure to keep the master cylinder full of fluid. With the bleeder valve open, pump the brake pedal. Watch the fluid squirting out. When air bubbles no longer appear, close the valve on the downward stroke of the pedal. Pump the pedal. If the pedal is solid, the air has been removed. If not, repeat the process.

Figure 15-9 shows a disassembled master cylinder. If fluid leaks out either around the pedal end of the cylinder into the boot or at the brake assembly, the master cylinder or brake cylinder is faulty.

Kits containing new rubber parts, a piston seal, boots, and valves may be available; if not, replace the entire unit. If the inside of the cylinder is corroded or scored, clean the surface with very fine wet or dry abrasive paper. If it does not clean up, replacing the cylinder is best. The cylinder can be honed to remove the scratches and scores, but it is generally best to replace the entire unit. Use alcohol to clean the internal parts; never use a petroleum-base parts cleaner, because oil damages the rubber parts. The system must be filled with new fluid and bled after being repaired. Lubricate all rod and cable connections.

16 TRANSMISSIONS

TRANSMISSION DESIGN
AUXILLARY TRANSMISSIONS
GEAR REDUCTION UNITS
POWER TAKE-OFFS (PTOs)
HYDRAULIC TRANSMISSIONS
TRANSMISSION CONSTRUCTION
GEAR REDUCTION UNIT CONSTRUCTION
POWER TAKE-OFF CONSTRUCTION
TRANSMISSION AND GEAR DRIVE REPAIR

The purpose of the transmission is to provide (1) a selection of gear ratios to meet varying load demands, (2) a neutral position so the engine can operate without the clutch being disengaged, and (3) to meet varying load demands, especially for starting from a standing position, the engine torque must be multiplied to varying degrees. Also essential is that the engine must be operable without the machine moving. Finally, many machines, especially tractors, have to be able to travel in reverse.

Numerous methods are used to achieve these purposes, particularly on small machines. Since these devices have been discussed in previous chapters, you will recognize them as you encounter them as torque multipliers:

1. The simplest mechanism for obtaining torque multiplication and neutral is to use a belt or belts with an idler pulley.

2. Another widely used mechanism that provides innumerable ratios is the variable sheave pulley system. But since this system

does not provide for reverse, the two-stroke cycle engines in some machines are simply operated in a reverse direction. The electric starting motor used on such an engine is designed to crank the engine for starting purposes, in a reverse direction—by moving a switch lever into reverse. The engine is designed to operate satisfactorily in either direction at the will of the operator.

3. Some tractors use a hydraulic transmission, which depends on oil pressure for its operation. The engine drives an oil pump to supply the pressure to operate a fluid motor, which then multiplies torque.

4. The most common transmission for tractor use is the sliding gear type. The difference in gear sizes between the drive gear and the driven gear brings about the desired gear ratio. When a small gear is meshed with a larger driven gear, the driven gear turns more slowly than the drive gear, but the torque output is greater. With a series of different-sized drive and driven gears, numerous ratios may be obtained. The different combinations are brought about by sliding the different-sized gears into mesh.

TRANSMISSION DESIGN

All transmissions are made up of different-sized gears. Some gears are in constant mesh with one another, and some are mounted on a splined or grooved shaft so they can be moved back and forth to engage other gears. Some transmissions use shift collars to lock different gears to a shaft to bring about a gear ratio change.

All revolving shafts are mounted on either bushings or antifriction bearings to reduce wear and friction, as well as to provide for replacement if too much free play develops.

Some gears which are known as "input" or "drive" gears, are connected to the engine crankshaft by means of a friction clutch. Other gears, known as the "output" or "driven" gears, are attached permanently to the rear axle assembly.

A gear shift mechanism, controlled by the operator, enables the operator to select the gear ratio that best meets operating requirements. The selecter (gear shift) must be designed so only one gear ratio can be in operation at any one time.

All transmissions are designed to include a neutral position, in which no connection is made between the drive (input) and driven (output) shafts. There are also provisions to lock the input shaft directly to the output shaft so there will be a direct drive between the engine and the rear axle assembly with no reduction. This is normally labeled "direct drive" or "high gear."

Reverse is obtained by placing into the gear train a third gear, which as an idler gear changes the direction of motion. When the reverse sliding gear is brought into mesh with the reverse output the reverse sliding gear is brought into mesh with the reverse output gear, this idler gear, located between the reverse drive gear and the reverse driven gear, causes the reverse driven gear to turn in the opposite direction from the drive gear.

The number of ratios (gears) depends on what the tractor is designed to do and on the loads that the machine is expected to handle. A number of tractors have a three-speed transmission: first gear (low), second gear (second), third gear (high), which is not a reduction gear but a direct drive. Included in all are a neutral and reverse. Figure 16-1 shows a cut-away illustration of a three-speed transmission, like the ones used in small garden and lawn tractors.

Some tractors have a four- or a five-speed transmission, which makes for more variables. Usually the aim is to get more power into the lower range, since the "high" gear—either third, forth, or fifth—is a direct drive with no reduction. Additional gears also provide for less difference in ratios between the different gears. Much depends on how the manufacturer designs the machine. The horsepower of the engine, the transmission ratios, and the ratio between the rear axle assembly and drive pinion all enter into the design.

Nevertheless, these factors have little or no bearing on the maintenance and servicing of the machine. Figure 16-2 is a cut-away view of a five-speed transmission, which provides more power ratios and greater flexibility. This transmission uses spur gears and bevel gears, along with a chain and sprocket to drive the different gear sets.

FIGURE 16-1
Three-speed transmission with sliding spur gears
Courtesy Engineering Products Co., Inc.

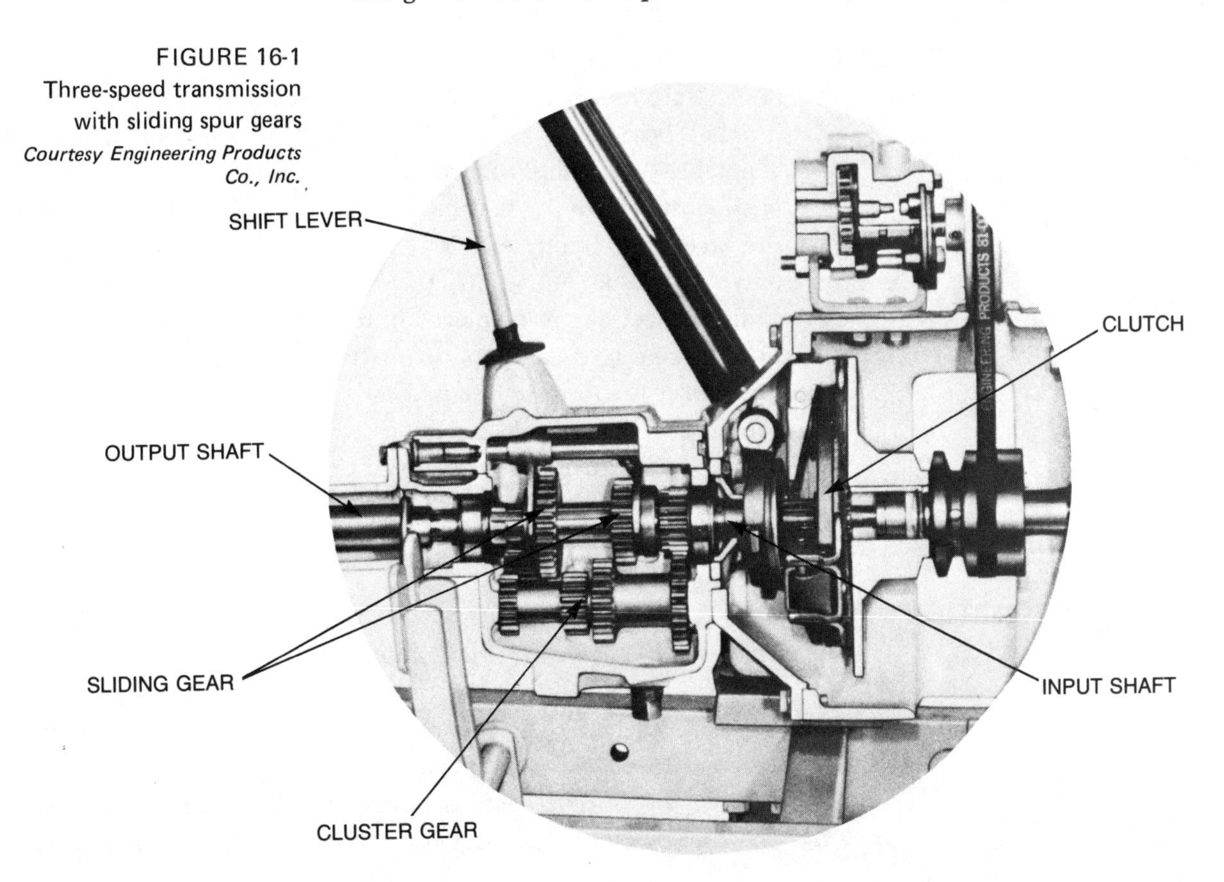

FIGURE 16-2
Five-speed manual-type transmission
Courtesy Deere & Company, Moline, Illinois

If properly used, transmissions cause little or no problems. Improper shifting can create problems by chipping gear teeth and placing undue strain on the gear teeth. Always be sure the throttle is released and the clutch fully depressed before attempting to shift gears. Often it is best to stop the machine before shifting gears. The transmission should always be properly lubricated with the specified lubricant at the required level, usually even with the bottom of the fill hole.

AUXILLARY TRANSMISSION

Some tractors are designed so that an auxillary transmission can be installed behind the regular transmission. This set-up makes a tractor adaptable to varying load conditions. If a three- or four-speed transmission takes care of the majority of demands, this is normally what the tractor is equipped with. But if it needs additional power for only occasional operation, then a second transmission, having an additional ratio, can be placed in tandem with the standard transmission. Generally it is simply a matter of replacing the present drive shaft with a shorter shaft and mounting the auxillary transmission behind the regular transmission case. Most auxillary transmissions are designed to provide for one additional reduction ratio and direct drive.

GEAR REDUCTION UNITS

To make an engine more adaptable for different uses, some manufacturers attach a gear reduction unit directly to the engine crankcase. A gear, attached to the crankshaft, is in constant mesh with a gear fastened to the engine output shaft. If more power is desired, the gear on the output shaft is larger than the gear on the crankshaft. The output shaft does not turn as fast as the crankshaft, but more power is delivered. If the load is light and higher speed is desired, the crankshaft gear is larger than the gear on the output shaft. This is a method of adapting the engine to a particular load requirement.

POWER TAKE-OFFS (PTOs)

In addition to propelling the machine, the engine may have to power auxillary equipment without moving the tractor or while moving it at different speeds than the tractor. Hence the transmission must be equipped with what are known as power take-offs (PTOs). A power take-off may be used to operate mowers, hole diggers, trenchers, a hydraulic pump to develop fluid pressure to operate lifts, and so on. The power take-off may develop more or less speed than the crankshaft by using gears, or it may be simply an extension of the crankshaft to which auxillary equipment may be attached.

Some power take-off units are located in front of the tractor and operate off the crankshaft. Others may operate off the transmission input shaft or the rear axle drive pinion. In any case, like the transmission and reduction unit, it involves gears, bearings, seals, and shafts. Some type of coupling is generally used for connection purposes.

HYDRAULIC TRANSMISSIONS

A few machines may be equipped with a hydraulic transmission system, which eliminates the need for a clutch and the shifting of gears. The system is made up of a variable displacement pump and a fixed displacement motor for the transmission of power. The engine-driven pump converts mechanical energy into hydraulic pressure while the motor converts hydraulic pressure into mechanical energy.

A piston-type pump, using a rotating shaft, imparts a reciprocating motion to several pistons, which then pump fluid under high pressure to the motor. The pistons are moved up and down in the cylinder body by the tilting action of a swash plate.

The system operates and is controlled by the rate of fluid flow, by the direction of fluid flow, and by fluid pressure. Controlling the rate of flow determines output speed. Control of the fluid flow direction determines reverse or forward. Control of fluid pressure results in control of the power output. The pump displacement, which governs volume and pressure, is controlled by a speed control lever arrangement.

The transmission is made up of many highly precision-fit parts. Should any service of the internal components be required, take extreme care to prevent damage. If a fluid leak develops or the transmission fails to operate with the proper linkage adjustment, it should be serviced by someone who is familiar with this type of transmission. Figure 16-3 shows a hydraulic transmission attached directly to the rear axle assembly. A flexible coupling is used between the transmission and engine drive shaft.

FIGURE 16-3
Hydraulic-type transmission
Courtesy Deere & Company, Moline, Illinois

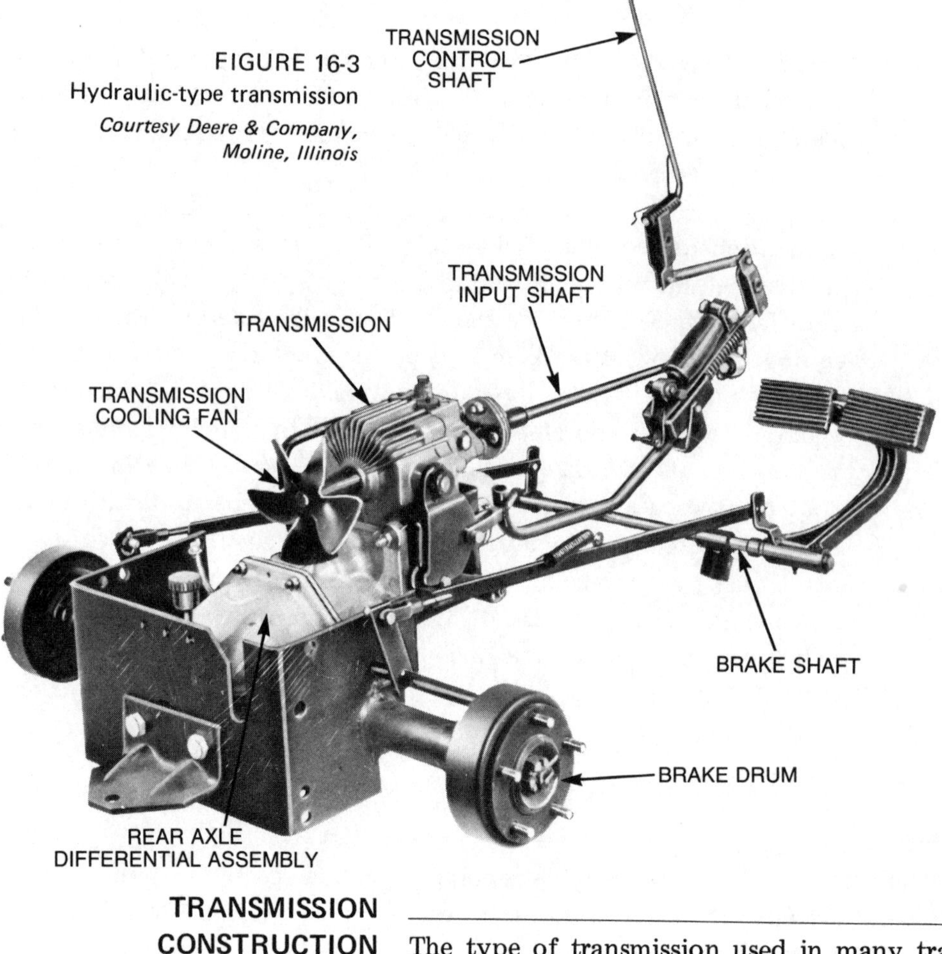

TRANSMISSION CONSTRUCTION

The type of transmission used in many tractors is generally much the same as the transmission used in older automobiles. The number of gear combinations used for reduction and the manner in which the gear change is brought about may differ in construction among various transmissions.

Figure 16-1 shows a typical three-speed tractor transmission using spur gears. Shifting into the various ratios is brought about by a shift lever operating two forks. When the shift lever is moved to the right, a gear change occurs when the lever is moved forward. Another gear change takes place when the lever is pulled back. The same thing takes place only with different gear changes when the lever is moved to the left and then moved forward and backward.

The clutch shaft drives the transmission input gear, which is in mesh with a reduction gear. The reduction gear is attached to the transmission countershaft gear. When a gear on the mainshaft is meshed (engaged) with a gear on the countershaft, the mainshaft is driven at a ratio determined by the number of teeth and size of the gears engaged.

The rear axle differential drive pinion is connected to the mainshaft of the transmission so that, whenever the mainshaft is turned, the drive pinion turns the rear axle assembly ring gear. Rounded corners on the gear teeth enable the sliding gear on the mainshaft to slide into mesh with less difficulty and with less chance of chipping the edges of the gear teeth.

Figure 16-4 shows a synchro-mesh unit where the gears are in constant mesh and turn freely on the shafts. A clutch collar is splined or keyed to the driving shaft. The collar is moved by the shifter fork, which is activated by the shift handle (lever) to provide the different ratios. The shift collar has notches on either side of the blocker ring. The notches engage openings in the gear when shifted into contact, thus locking the gear onto the shaft so it drives the gear attached to the output shaft. In this particular illustration, the shift collar is used to provide forward and reverse.

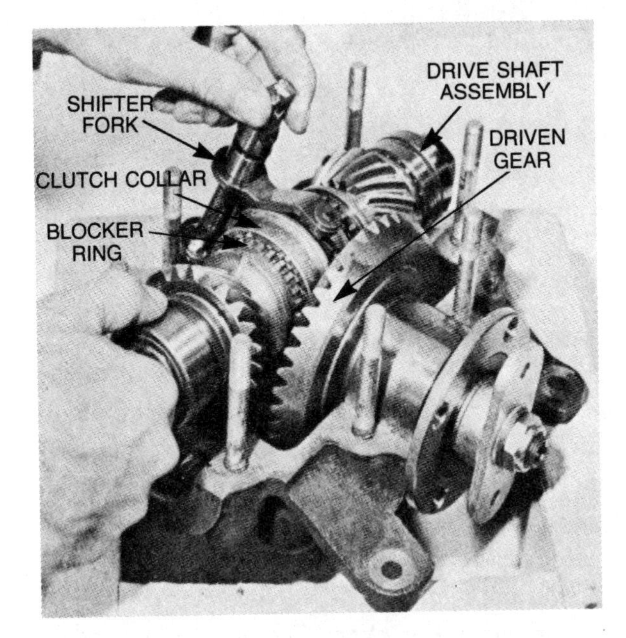

FIGURE 16-4
Syncro-mesh unit and shifter fork used for forward and reverse
Courtesy Harley-Davidson Motor Co., Inc. Subsidiary of AMF, Inc.

GEAR REDUCTION UNIT CONSTRUCTION

Some tractor manufacturers make available an additional gear ratio —in some cases labeled a "creeper drive"—in the form of an auxillary transmission. When engaged, it lowers the speed of the machine considerably but greatly increases the power delivered to the wheels. Since the unit is generally installed at the front of the transmission when the transmission is attached directly to the rear axle assembly, it can be added with few alterations other than a shorter drive shaft or clutch shaft.

The unit consists of a planetary gear train, a shifter collar, and a direct drive coupling. It is controlled by a shift lever operating a shifter collar and yoke. With the unit in direct drive, the input shaft and the driven shaft are coupled together and no gear reduction takes place. When shifted into "creeper drive," the shifter collar moves the sun gear so the ring gear drives the driven shaft at a reduced speed but increased torque. Figure 16-5 shows a planetary gear train.

Some small engines have mounted directly on the engine a gear reducer, which is driven by the crankshaft and which generally uses the same lubrication system as the engine. A gear reducer can be obtained with different reduction ratios so as to tailor the engine to

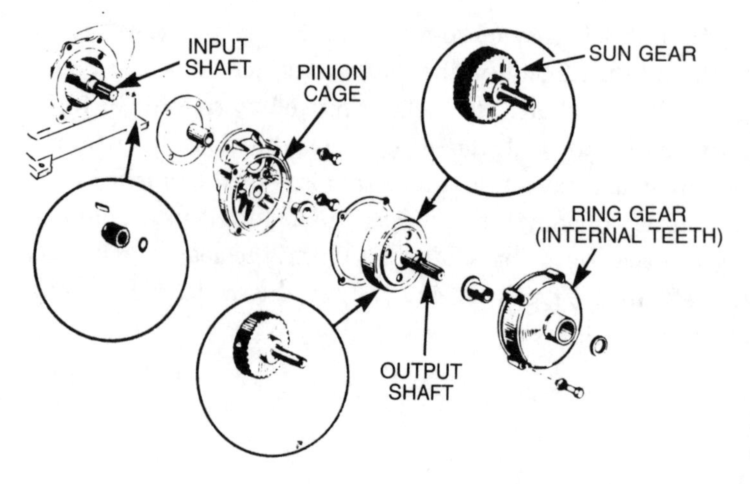

FIGURE 16-5
Planetary gear reduction unit
Courtesy Clinton Engines Corporation

specific load requirements. It is made up of a planetary gear train consisting of a sun gear, planetary pinions mounted on a cage, and a ring gear with internal teeth. All the gears are in constant mesh so no shifting takes place. There is little chance for wear. Figure 16-5 shows a typical gear reduction unit.

POWER TAKE-OFF CONSTRUCTION

A power take-off, used to drive additional equipment, may be installed on a self-propelled machine. It can take the form of a pulley to drive a belt or simply a shaft extending from the crankshaft, transmission, or rear axle assembly. The shaft or pulley may be used to drive a piece of equipment. Instead of depending on the movement of the machine, a piece of equipment can be operated off the tractor when the tractor is not moving. For example, the power take-off may be used to drive a mower's cutting blade, the blower of a snow blower, or a hydraulic pump to provide hydraulic pressure to operate a lift, blade, or hoist.

The simplest power take-off is an extension of the crankshaft beyond the front of the engine, which mounts a pulley or a coupling to which a driven shaft can be attached.

Figure 16-6 shows a power take-off shaft that is perpendicular to the crankshaft. This shaft, driven by the camshaft, turns only once every 8½ revolutions of the crankshaft due to the number of teeth on the gear and the gear size. This ratio develops power but not much speed. Some power take-off assemblies are mounted on the front end of the crankshaft (Figure 16-7).

Some may use a brake and clutch system, which permits the pulley to be disconnected from the crankshaft without interfering with the engine operation. The clutch and brake are controlled by a lever. The brake is a safety feature to permit control of the machine being driven by the power take-off.

The power take-off may also be located on the differential housing and powered by the drive pinion, or it may be attached to the transmission input shaft. If powered by the drive pinion, the

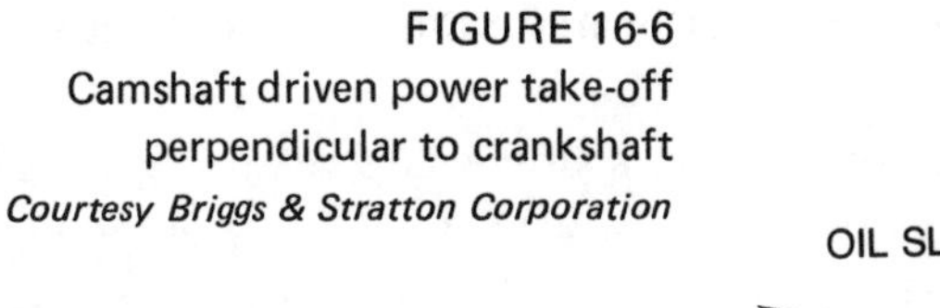

FIGURE 16-6
Camshaft driven power take-off perpendicular to crankshaft
Courtesy Briggs & Stratton Corporation

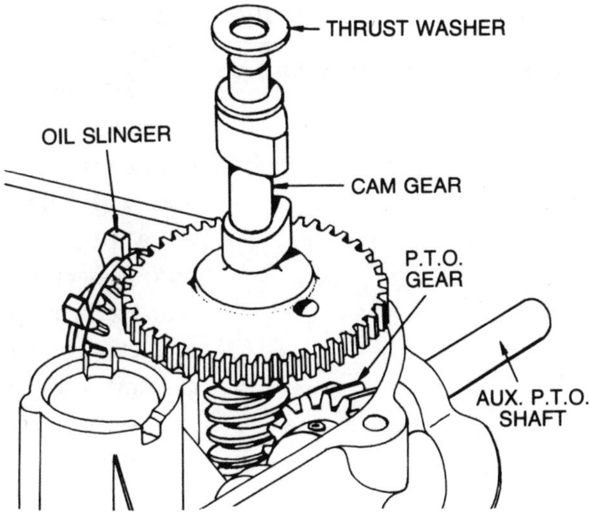

FIGURE 16-7
Power take-off mounted on front end of crankshaft
Courtesy Engineering Products Co., Inc.

power take-off would be in operation only when the machine is moving. If the power take-off is driven by the transmission input shaft, power is applied to the unit whenever the engine is running and the clutch is engaged.

TRANSMISSION AND GEAR DRIVE REPAIR

Transmissions, gear reduction assemblies, and power take-off units are made up of gears, shafts, bearings, bushings, oil seals, and spacers (washers). With the necessary maintenance—such as keeping the proper lubrication level, shifting without jamming gears (proper use of the clutch), and making sure all linkages are in adjustment—transmissions and other geared units give satisfactory service over a long period.

Malfunctions in a unit using gears is generally noticeable by its noisy operation. If a gear tooth is badly chipped or missing, a knocking prevails as the gears are used. This noise may change or stop if a gear with chipped teeth is shifted out of mesh, whereas a whining noise that varies with engine speed can indicate a faulty bearing. If you suspect the transmission or gear unit of having either chipped gear teeth or faulty bearings, drain the lubricant into a pan. If flakes or metal chips drain out with the lubricant, you have to disassemble the unit to locate and repair the problem.

If the case has a removable cover on the side, or if the shifter fork cover assembly can be removed, then taking off the cover or shift lever assembly should permit you to inspect the condition of the gears and bearings. Gear wear is usually obvious. Bearing wear can be determined by the amount of free play you get when you try to move the gears that are mounted on bearings. End play, which is controlled by washers and/or spacers, can be determined by the amount of movement in a shaft or gear cluster.

After the trouble has been isolated, you must determine the amount of work needed to make the repairs. Gear reduction and power take-off units can generally be removed from the engine or other units with very little work. But to service the transmissions in many tractors, you have to split them. The degree of disassembly varies from machine to machine. A certain amount of sheet metal—such as fenders, foot platforms, as well as brake control rods and cables—may have to be removed. The front section of the machine has to be supported before the rear axle assembly can be removed. Couplings and other connections need to be taken out before unit is separated. Usually you can see what has to be removed.

When disassembling a transmission or gear reduction assembly, as well as any other components with numerous parts, note the relative locations of parts and components before their removal. Use labels and different pans so as to make sure everything is reinstalled in its exactly original location. Keep a close account of the location and number of washers; they usually control end play.

Clean all parts thoroughly in a grease solvent. Inspect antifriction bearings for roughness and wear. Inspect bushings and shafts for excessive clearance. Inspect the gears for wear and chipped teeth. Use new gaskets and oil seals when reassembling. Bearings and bushings should have no free play but must not bind or drag. Washers (shims) of different thicknesses are usually available to adjust end play. End play should be kept to a minimum; bearings and bushings must turn freely but have no play or drag.

Always refill the assembly with new lubricant to the correct level.

17 FINAL DRIVES

PULLEYS
SPROCKETS
GEAR REDUCTION AND DIFFERENTIALS
LOCATING DRIVING AXLE PROBLEMS

A small gasoline engine generally does not have enough starting torque (turning power) to initially move a heavy load. Furthermore, most small engines operate better at high speeds. For this reason in most installations the speed that the engine produces at the crankshaft must be utilized to develop power and to reduce speed. So, with few exceptions, some type of reduction mechanism is placed at the final drive for most pieces of equipment powered by small gasoline engines—especially if they are propelled by the engines. The common exception is the push-type lawn mower, where the cutting blade is attached directly to the engine crankshaft.

Different methods are used to obtain the necessary driving force. Selective torque multiplication may be provided either by means of a transmission or variable drive sheave or by means of a connection between the clutch and drive axle, both of which were previously discussed in the chapters on "Coupling Devices" and "Transmissions." In most installations, however, a final drive provides the additional reduction, commonly through the use of (1) pulleys and belts and (2) sprockets with chain (3) and gearing.

PULLEYS

The simplest method of obtaining a reduction in speed and an increase in power is through the use of different-sized pulleys connected by belts. Figure 17-1 shows a small pulley, attached to the starter, which drives a larger pulley on the engine crankshaft. This pulley, in turn, is used to crank the engine. When the engine starts, the belt tightener (clutch) automatically disengages the starting

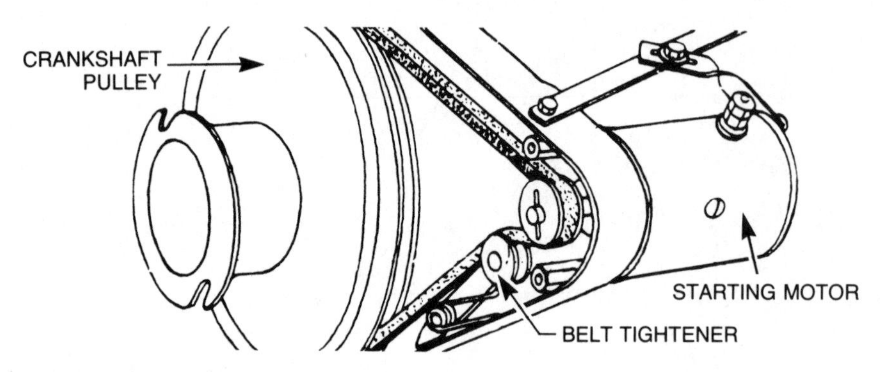

FIGURE 17-1
Belt drive starting motor
Courtesy Briggs & Stratton Corporation

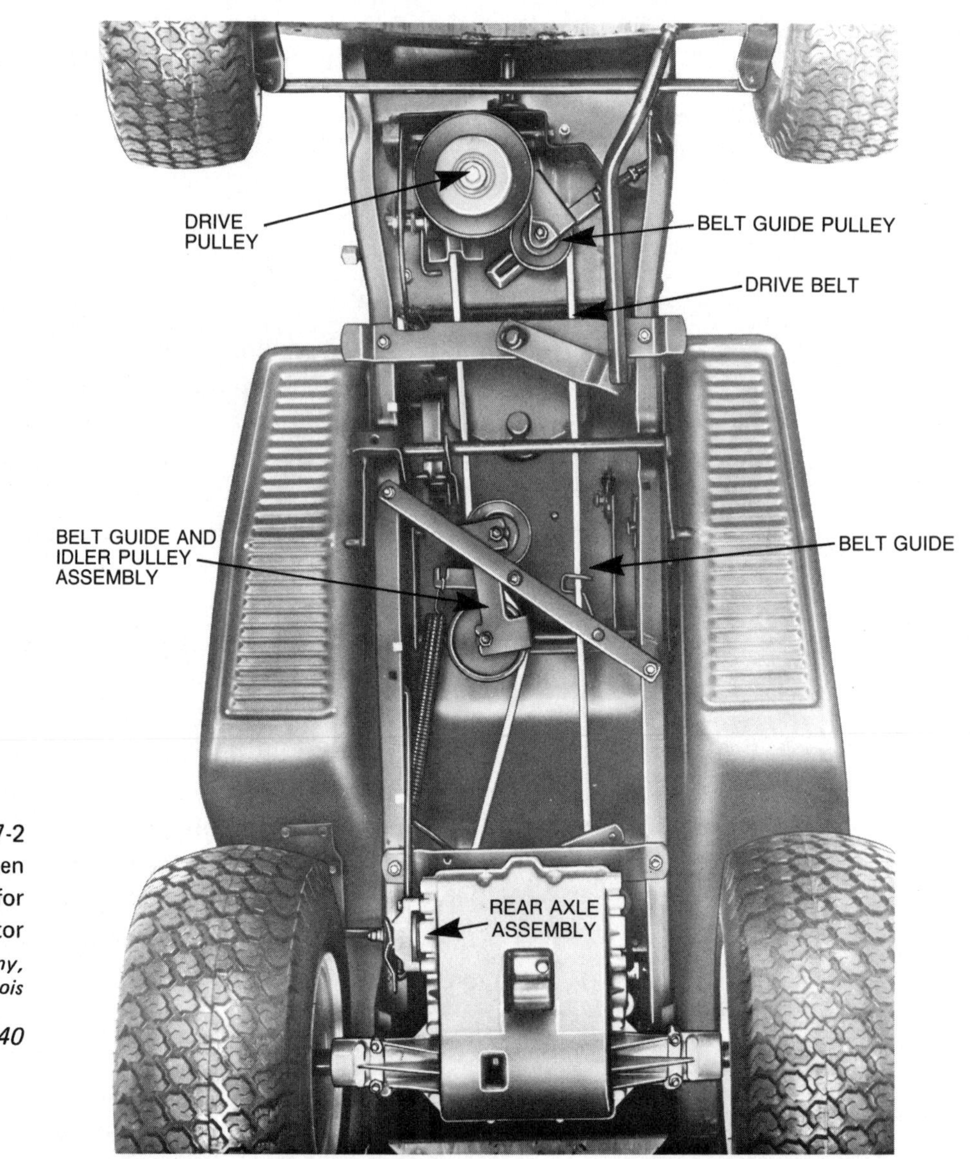

FIGURE 17-2
Belt and pulley drive between engine and rear axle for lawn and garden tractor
Courtesy Deere & Company, Moline, Illinois

motor from the engine. The driving force is applied to the sides of this vee-type pulley. On some installations a flat pulley may be used, but a flat belt must then be used with a flat pulley.

Whatever ratio is necessary to produce the desired results may be obtained by varying pulley sizes, as dictated by the horsepower of the engine and the load to be imposed. If the drive pulley—a crankshaft pulley as an example—is the same size as the driven pulley, there is no change in power or speed between the two units. If the drive pulley is larger than the driven pulley, the driven pulley revolves faster than the drive pulley but with less power. If the drive pulley is smaller than the driven pulley, the driven pulley turns more slowly but produces additional torque.

Figure 17-2 shows the pulley-and-belt arrangement used on a lawn or garden tractor to drive the rear axle assembly. A pulley attached directly to the engine crankshaft drives a belt. The belt turns a pulley located on the final drive to power the rear wheels. A pulley mounted on a slotted bracket serves as a belt guide. A combination of idler pulley and belt guide is used to engage and disengage the engine from the final drive. A coil spring maintains tension on the idler pulley.

SPROCKETS

Another arrangement that serves the same purpose involves sprockets and a chain. The sprocket serves the same function as the pulley, and the chain takes the place of the belt. Figure 17-3 shows a chain with a chain tensioner, which helps to maintain the proper tension on the chain as wear takes place.

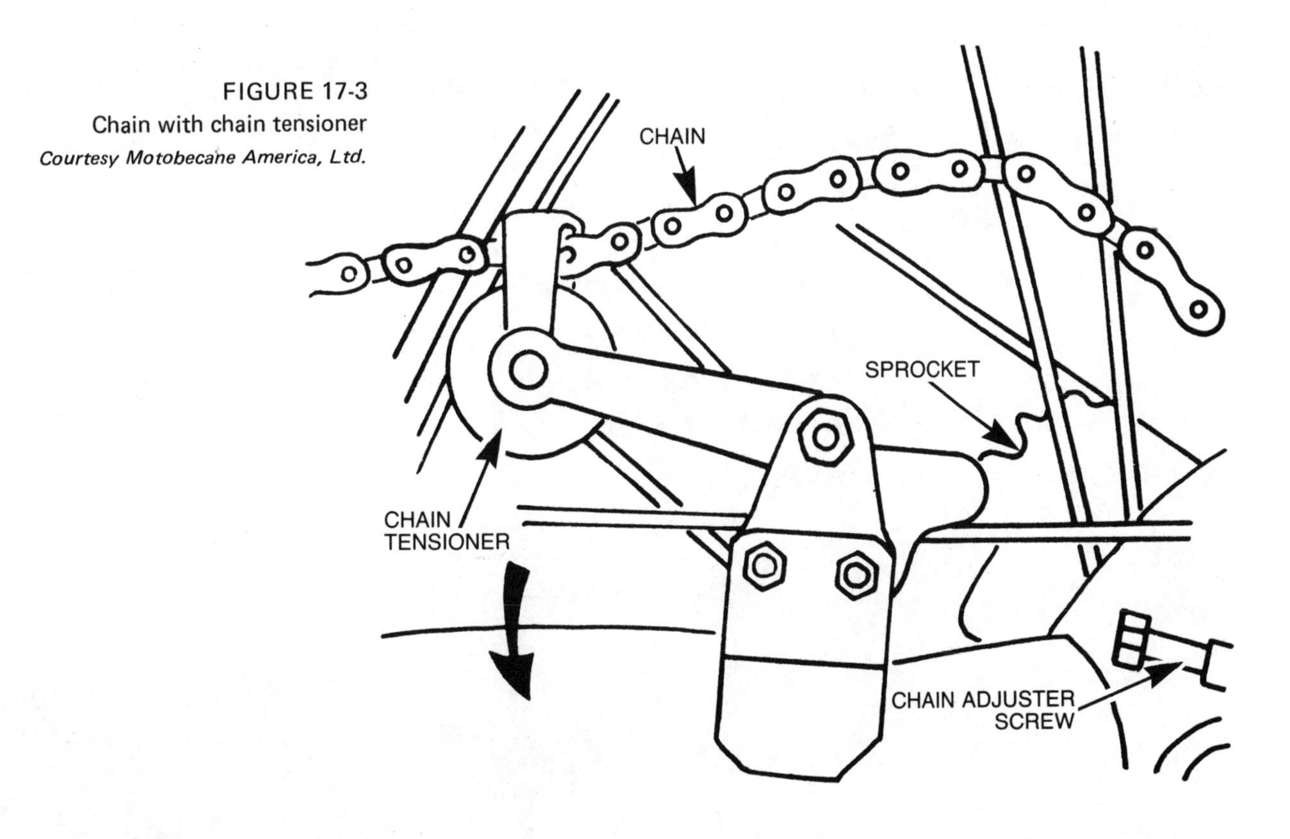

FIGURE 17-3
Chain with chain tensioner
Courtesy Motobecane America, Ltd.

GEAR REDUCTION AND DIFFERENTIALS

Many of the rear axle assemblies in riding tractors are of the same general type as those used in automobiles. The rear axle assembly has three basic functions to perform on these machines: (1) It provides a gear reduction ratio for the driving wheels that propel the machine. (2) It contains a differential assembly that permits one wheel to turn faster than the other when rounding a corner. (3) It supports part of the machine's weight.

The gear reduction ratio is determined by the number of teeth on the drive pinion and ring gear, which do the same thing as the different-sized pulleys. The ratio also depends on the design of the machine and its use: the higher the ratio, the more power and the less speed.

The drive pinion may be connected to the engine by a belt or shaft, or it may be connected directly to the transmission. The differential assembly consists of differential pinions mounted in and turning with the differential housing. These pinions are in constant mesh with the side gears, which are splined to the shafts. Figure 17-4 shows complete differential assembly with the cover removed. The axle shafts extend out from either side of the differential assembly, and they are contained in a housing which bears part of the machine's weight.

The drive pinion, differential housing, and outer ends of the axle shafts are generally mounted on ball or roller bearings. When correctly adjusted and lubricated, the assembly performs satisfactorily over a long period with little or no service. The ring gear and drive pinion must maintain the correct mesh (tooth clearance) to prevent tooth wear and noise. The gear mesh, although adjustable,

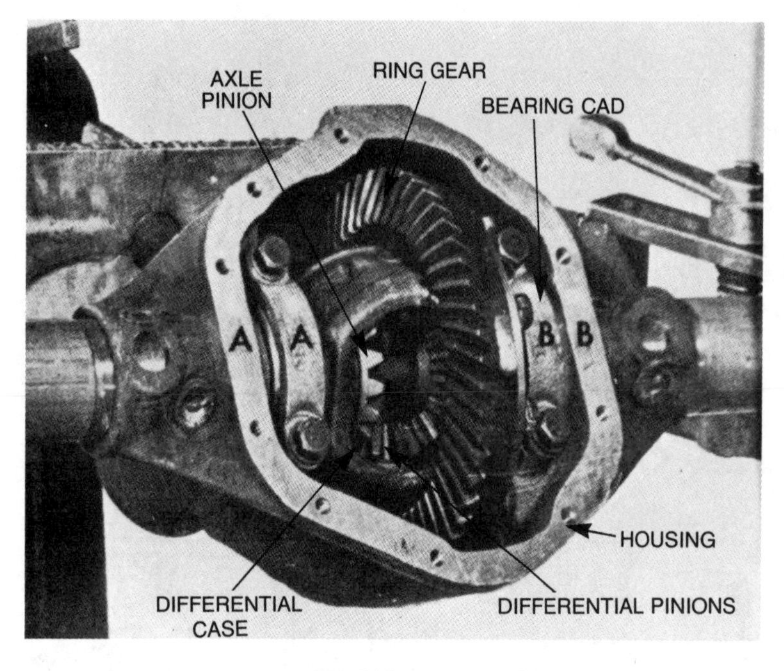

FIGURE 17-4
Rear axle differential assembly
Courtesy Harley-Davidson Motor Co., Inc. Subsidiary of AMF, Inc.

should remain fixed as long as there is no bearing wear or misadjustment. A properly adjusted rear axle assembly should operate quietly without an excessive amount of backlash, which is the free play (clearance) between the gear teeth.

Some other tractors use a differential like the automotive type just described, but, to obtain more torque, they use an additional set of gears at the rear wheels as a final reduction drive. A gear located near the end of each axle shaft is in constant mesh with a large spur gear that is keyed to the rear driving wheels. The gears are enclosed in a housing (Figure 17-5). With an approximate reduction of 5 to 1 in the regular differential assembly, plus a reduction of about 10 to 1 in the final drive gear, the power unit can provide a lot of power at the driving wheels.

Another type of installation uses the same basic components, but it is constructed so as to reduce the size of the differential housing. It does not enclose the axle shaft or ring gear and pinion. The unit is supported by bearings in a housing attached to the frame at each wheel. A sprocket bolted to the differential takes the place of the ring gear. A small sprocket on the transmission output shaft drives the differential sprocket by means of a chain. The two sprockets serve the purpose of the ring gear and pinion. The difference in size between the drive sprocket (transmission) and the driven sprocket (differential) determines the ratio. The bearing housing mounting bolt holes in the frame are slotted at the top, to enable you to move the axle assembly back for proper chain adjustment. Adjust the bearing housing until the slack is out of the chain, but the chain must not be taut.

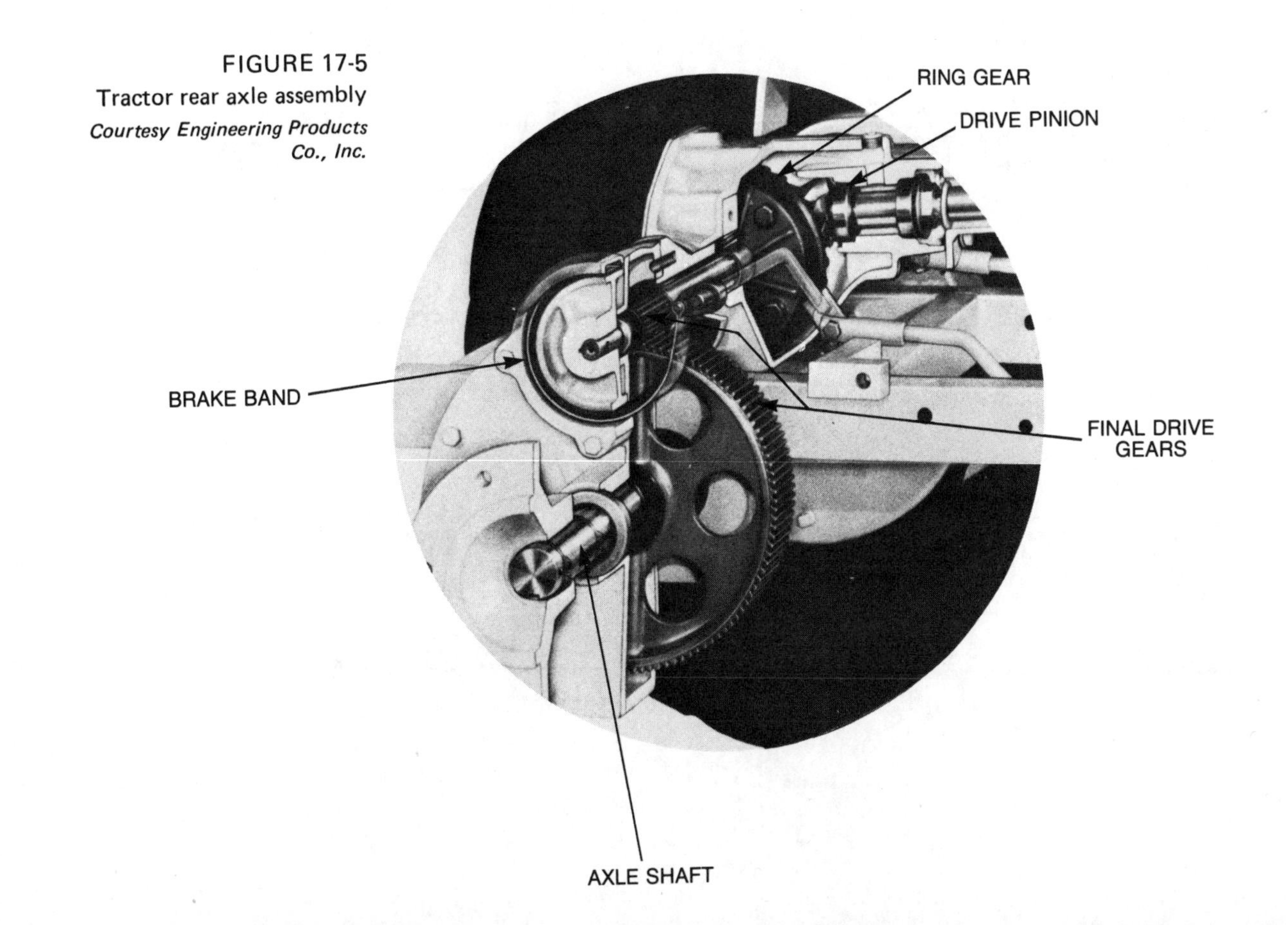

FIGURE 17-5
Tractor rear axle assembly
Courtesy Engineering Products Co., Inc.

Figure 17-6 shows a combination of transmission, rear axle, and differential assembly, which is sometimes referred to as a "trans axle." Selective sliding spur gears in the transmission brings about the different ratios. A spline-type gear, located in the bottom of the transmission, serves as the rear axle differential drive pinion and is in mesh with the ring gear. The ring gear takes the form of a spur gear. The assembly is serviced as one unit.

The different brands of soil tillers or garden tillers are all powered by small single cylinder gasoline engines. Most tillers have two driving wheels, along with different tine arrangements that, in addition to breaking up the soil by rotating, help to propel the machine. A main shaft, driven by a belt, has a worm gear on either end, one is used to drive the tiller tine holders and the other to drive the wheels. There is no differential action, and the unit may be referred to as a "transmission." Both driving wheels turn at the same

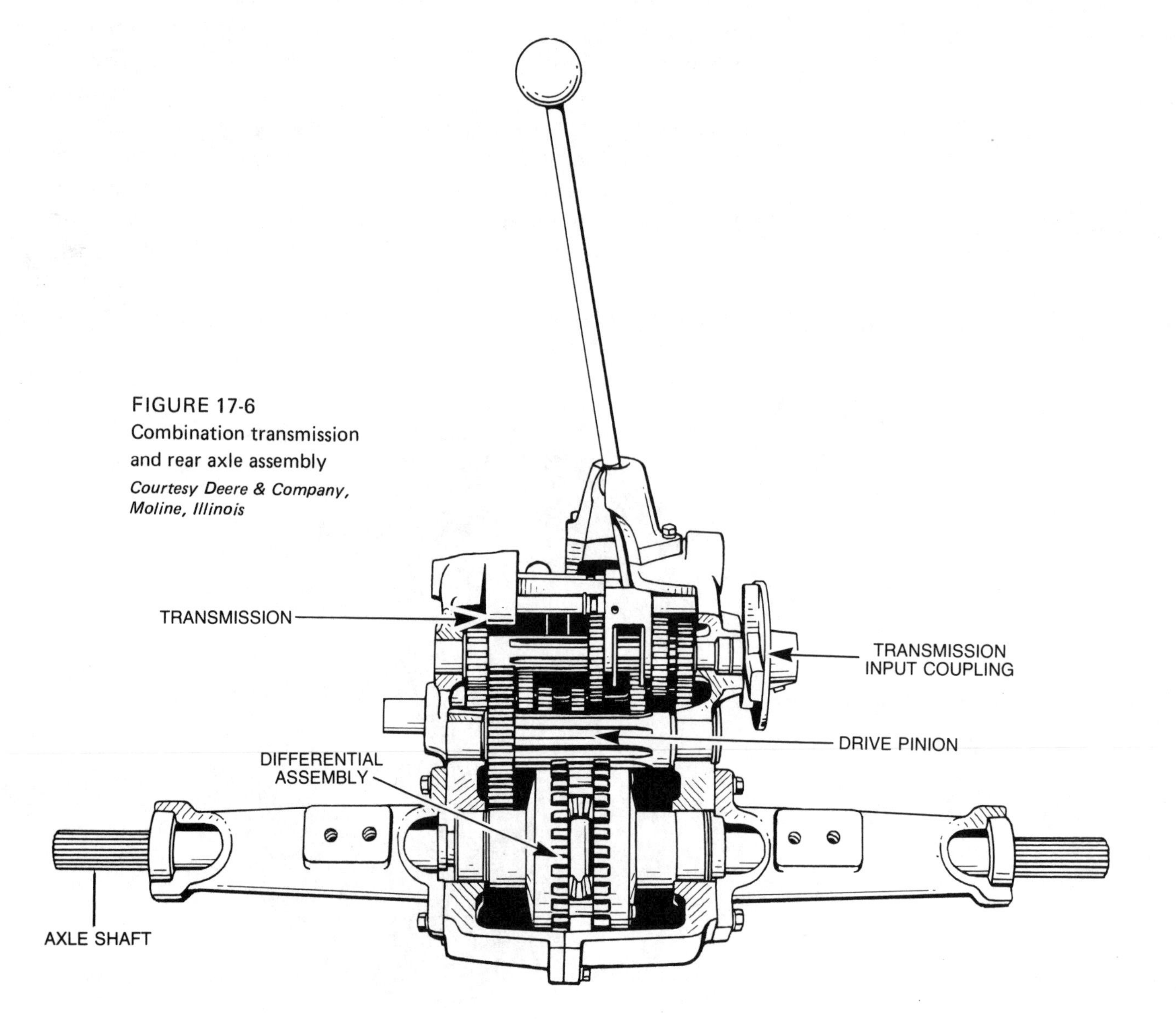

FIGURE 17-6
Combination transmission and rear axle assembly
Courtesy Deere & Company, Moline, Illinois

speed, regardless of load. The driven worm gears are in mesh with gears to drive the tiller tines and wheels. A sliding clutch permits the operator to select a low or high tiller speed. The driving torque from the worm gear is transmitted through spur gears to the shaft, which powers the wheels. Provisions are made for changing drive pulleys for increased or reduced speed.

LOCATING DRIVING AXLE PROBLEMS

Locating rear axle problems is not a complicated process. Raise the driving axle assembly with a jack, and let it down on stands or blocks so that both wheels are free to turn. Be sure to chock the other wheels so the machine cannot fall off the blocks. Start the engine and shift into drive. Listen for noises. A constant distinct knock or loud click, which changes as engine speed is changed, can indicate a broken or badly chipped gear tooth on the ring gear or pinion.

To isolate the noise, you need a metal rod about 3/8 inch in diameter and 12 to 18 inches long, or a screwdriver with metal through the handle. Place one end of this "sounding rod" or screwdriver against your ear and the other end against various places on the housing. The tool amplifies the noise, making it easier to locate where the noise is coming from.

1. Vary the speed of the engine. A whine or growl that occurs on acceleration and/or deceleration can indicate ring gear and pinion adjustment or wear.

2. A steady noise may indicate a bad bearing. By using a sounding rod, you should be able to determine whether an outer wheel bearing or some other bearing is at fault.

3. If the noise occurs only when one wheel is held, it would indicate a problem in the differential gear assembly, possibly a broken or chipped gear tooth.

4. If one wheel does not turn, hold the wheel that has been turning. If the opposite wheel does not turn, the axle shaft is broken.

Repairing a driving axle assembly, particularly on a larger tractor, generally requires splitting the tractor at the transmission. This is heavy work, and you need jacks and blocking material to support the two halves. You also need special tools, such as a bearing remover, dial gauge, or other fixtures to get the proper bearing pre-load, pinion depth, and gear backlash. The repair requirements should not deter you, however, from determining the trouble.

Most differential housings have a removable cover, which you can remove once you drain the fluid from the housing. Then you can inspect most of the components that make up the differential assembly, as well as the ring gear and drive pinion. Metal flakes, broken gear teeth, or other pieces of metal in the bottom of the housing indicate trouble and the need to disassemble the unit. By moving the ring gear back and forth by hand a little bit, you can feel how much backlash is present. The amount of free play between the teeth on the ring gear and the drive pinion should be approximately 0.003 to 0.005 inch, and you can accurately measure it with a dial gauge. By applying red lead or prussian blue to the teeth on the ring gear and then turning the gear, you can check the tooth bearing pattern on the ring gear and pinion. Check the tooth pattern. Correct tooth contact and gear adjustment are indicated by a wear pattern in which the red lead or prussian blue is rubbed off in the center of the tooth.

If you have to remove an axle shaft, on some installations you must remove a C-type washer from the inner end of the axle shaft. On other installations, remove the rear wheels and then remove the axle shaft bearing retainer. The axle shaft can now be pulled out of the housing, thus permitting you to service the wheel bearing and to replace the oil seal if necessary. Figure 17-7 shows an exploded view of a rear axle assembly.

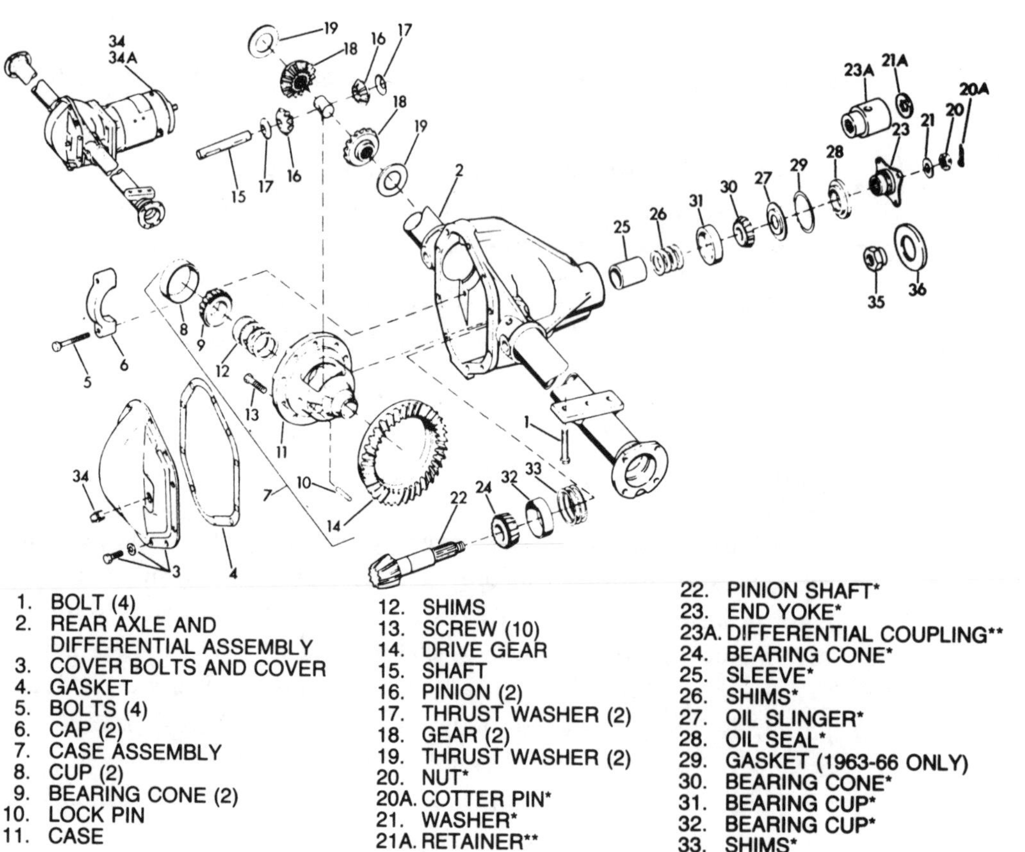

FIGURE 17-7
Rear axle assembly completely disassembled

Courtesy Harley-Davidson Motor Co., Inc. Subsidiary of AMF, Inc.

A new gasket should always be used when reinstalling the differential cover. Be sure all the old gasket material is removed before installing the new gasket. Always fill the housing with new fluid to the proper level, which is the bottom of the fill hole.

All gears, bushings, bearings, and shafts are inspected in much the same way regardless of where they are used. To check gears for signs of wear, look for any chipping on the tooth surface, roughness, case hardening worn through, or discoloration. The gear must be replaced if any of these conditions exist. Ring gears and pinions must be replaced in sets. Check bushings for roughness and out-of-round. If any chipping or uneveness is evident on the rollers of a bearing, or if the bearing does not turn smoothly, then replace the entire bearing assembly including the cup. If the splines or key way on the shafts are worn, replace the shaft.

All gears must be adjusted so there is very little free play, approximately 0.003 to 0.005 inch. Figure 17-8 shows the use of a dial gauge to check the free play known as "running clearance," between the teeth on a ring gear and pinion.

All bearings must be adjusted so there is no free play, but they must not create pressure on the bearing cup.

FIGURE 17-8
Checking ring gear backlash with a dial gauge
Courtesy Harley-Davidson Motor Co., Inc. Subsidiary of AMF, Inc.

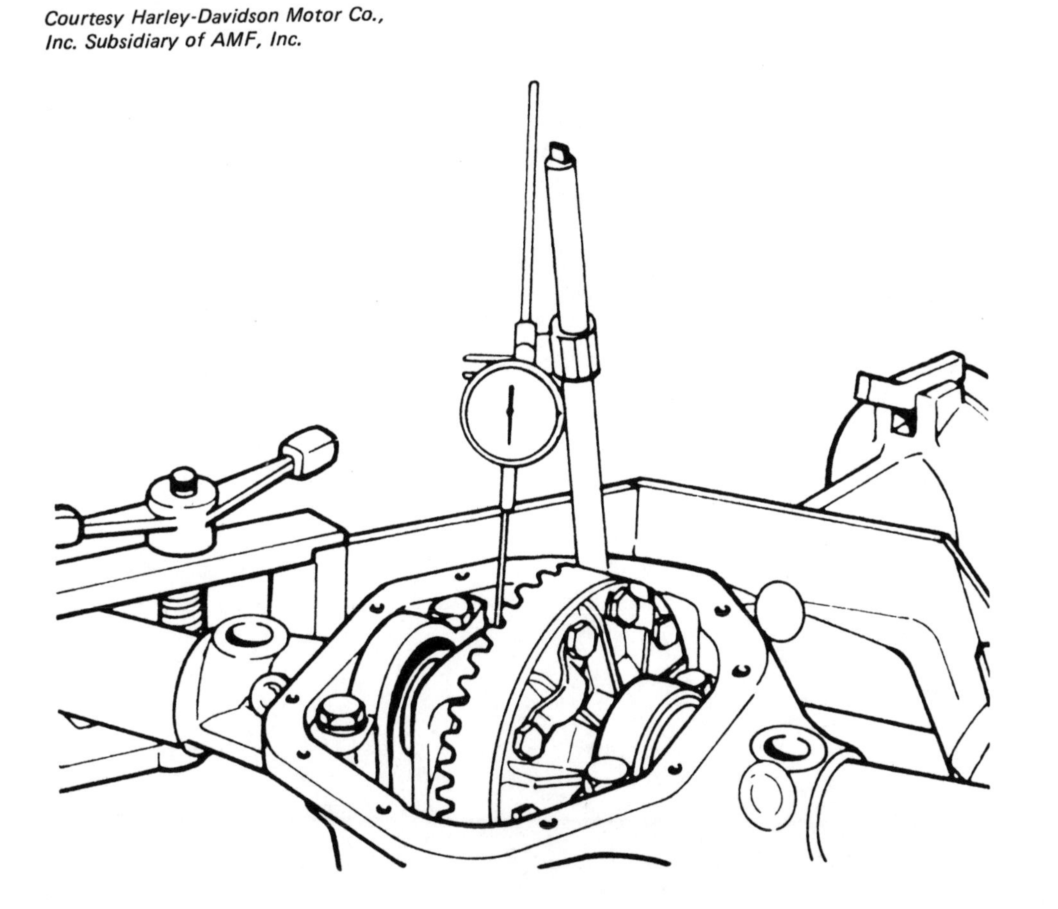

18 PREVENTATIVE MAINTENANCE

MAINTENANCE MEASURES
SAFETY PRECAUTIONS
STORAGE STEPS

Maintenance, a preventative operation, is performed to reduce wear, to prolong the life of the equipment, and to make for more efficient operation. While doing preventative maintenance work, you may also discover other problems such as loose bolts, wear in a component, or the start of a malfunction. If you discover these conditions before they become serious, many times you can take care of problems before they begin to affect operation and cause trouble or wear in other units.

MAINTENANCE MEASURES

Service schedules vary somewhat among manufacturers. The owner's manual generally gives the specifications and maintenance schedule for a piece of equipment. When a specific schedule is not available, use the general information in this section as a guide. Of course, you learn the details for performing the different maintenance operations from the text material throughout the book under the appropriate headings.

In every case two factors have a bearing on the maintenance schedule: (1) the type of use to which the machine is being put and (2) the operating conditions, such as dust and dirt. When the machine is regularly worked at its capacity under heavy loads, and/or when the working conditions are dusty or dirty, many of the maintenance operations should be performed more regularly: checking the oil, servicing the air cleaner, lubricating the machine, and so on.

1. The oil level in the crankcase should be checked daily or at least each time fuel is added. The oil should be changed after every 25 hours of operation. Under extremely dusty conditions change the oil even more often.

2. The air cleaner should be checked and cleaned every 25 hours. The dry element should be replaced after every 200 hours of normal operation.

3. The fuel sediment bowl and filter element, if used, should be cleaned every 100 hours. With a glass sediment bowl, you can see when foreign matter accumulates in the bowl.

4. The spark plug should be removed, cleaned, and regapped every 25 hours. The plug should be replaced if the electrodes show signs of eroding and/or the insulator is cracked or blistered.

5. Inspect the condition of the ignition breaker points every 150 hours for pitting and gap setting. Check ignition timing, and regap the points after 50 hours. The points may be pitted or eroded to the extent that they should be replaced.

6. Check the belt tension and condition of the drive belts, as well as the starter/generator and other belts, every 50 hours. Replace the belts that are frayed, cracked, or stretched to where they can no longer be adjusted.

7. Check the external air cleaner screen—generally located on the housing for the hand-operated starter—along with the cooling fins and external block, especially around the oil fill area. Check for dirt, grass, and other foreign matter.

8. Every 500 hours, check the tightness of all the fasteners, especially mounting bolts, to keep damaging vibrations at a minimum. Make sure the fuel tank cap vent is free from foreign matter.

9. In two-stroke cycle engines, remove and inspect the muffler and exhaust system every 50 hours. Any build-up of carbon in the exhaust port, muffler, or cylinder affects engine efficiency. The carbon can be scraped loose and then removed by using a blunt putty knife or other scraper. Be careful not to mar the surfaces.

10. If the engine is equipped with a battery, check the electrolyte level about every two months if the engine is used regularly. Make sure the hold-down brackets are tight. Clean the battery terminals if they are corroded, as well as the top of the battery if it is dirty. To clean battery posts and terminals, as well as the battery top and carrier, use a solution of one tablespoon of baking soda to a cup of water. But make sure no soda water gets into the battery. The battery terminals must be kept clean and tight. A light coating of petroleum jelly or light grease helps to keep the terminals clean.

11. In tractors, tillers, mowers, and other pieces of equipment that have units lubricated through grease fittings, lubricate the parts every 50 hours. These parts include the steering linkage, suspension system, wheel bearings, and various shafts. Also check the fluid level in the transmission and rear axle. For grease fittings, use a regular chassis lube, such as the type used for an automobile. To force grease through the fittings, you need a grease gun, which you may purchase at an automobile parts and accessory store.

SAFETY PRECAUTIONS

1. When operating any power equipment, take care to prevent and to avoid accidents. Always use caution around any moving equipment. Keep your hands, feet, and loose clothing out of the way of moving blades, belts, shafts, and other parts.

2. Do not add fuel while the engine is running, and if possible, allow the engine to cool before adding any. Spilled fuel can ignite if it comes into contact with hot engine parts.

3. Always disconnect the spark plug cable from the spark plug to prevent the engine from starting, should the engine be cranked while working on the engine or equipment. Be especially aware of this rule when working on the blade of a rotary mower.

4. Disengage all equipment from the engine before starting. Make sure that the transmission is in neutral and that the power take-off is disengaged.

5. Make sure all safety guards are in place on the engine and equipment before operating.

6. Do not change the setting of the governor, which establishes safe operating limits.

STORAGE STEPS

If the machine—such as a mower at the end of the mowing season or a snow blower or snowmobile in the spring—is not going to be used for several months, you should do a number of things before putting the maching into storage. A little care in storage keeps the machine in good condition while not being used.

1. When an engine sits over a period of time without being used, oil can drain off the cylinder walls and other moving parts, allowing rust to occur. To prepare for storage, run the engine until it is warm. Then completely drain all the oil from the crankcase. Refill it with new oil, and run it for a few minutes to circulate the new oil. Then leave the new oil in the crankcase.

2. Fuel left in the tank becomes stale and no longer vaporizes when used at a later date. Also, fuel left in the carburetor may evaporate, leaving a gummy residue that plugs up the jets and other small openings. The same thing happens to the fuel pump. Drain the fuel from the fuel tank. For both four-stroke and two-stroke cycle engines, start the engine, and run it until it stops for the lack of fuel.

3. On two-stroke cycle engines, remove the exhaust pipe and muffler. Clean the carbon out of the exhaust port and on top of the piston and cylinder head. Scrape loose whatever carbon you can get at with a blunt instrument, such as a putty knife, old table knife, or screwdriver. Do not mar the metal parts. Loose carbon on top of the piston can be removed by taking out the spark plug and rotating the engine crankshaft.

4. After the spark plug is removed, pour a tablespoon of oil into the spark plug hole. Turn the engine over a number of times to make sure the cylinder wall, rings, piston, and valves are coated with oil. Clean and regap the spark plug, if it is to be reused, or obtain a new one if needed. (Always check and set the gap on a new plug.) Install the spark plug.

5. Service the air cleaner or replace it with a new one if necessary.

6. Inspect the cooling fins on the engine cylinder and cylinder head for dirt and grass build-up. Carefully scrape away the dirt and/or grass. To completely clean out an excessive amount, you may have to remove the engine shroud.

7. In a mower, clean the entire mower section with soap and water. Using a putty knife, break loose and scrape off the caked material on the mower housing and around the blade. If the mower blade needs sharpening, now is a good time to do it—or at least have it done before the grass cutting season begins.

8. Cover the exhaust pipe or muffler opening with a tin can or tape to prevent dirt and dust from entering the engine.

9. In a snowmobile, the drive belt should be removed and the faces of the clutch oiled to prevent rusting. The tension on the track

should be reduced and the machine blocked up. The track should not rest on the ground or concrete surface.

10. Put a light film of oil on surfaces that might rust, such as cutter blades, tiller blades, and so on.

11. Lubricate the steering linkage, control cables, and other moving parts.

12. If a chain saw is to be stored for any extended length of time, remove the chain and store it in a container, covered with oil.

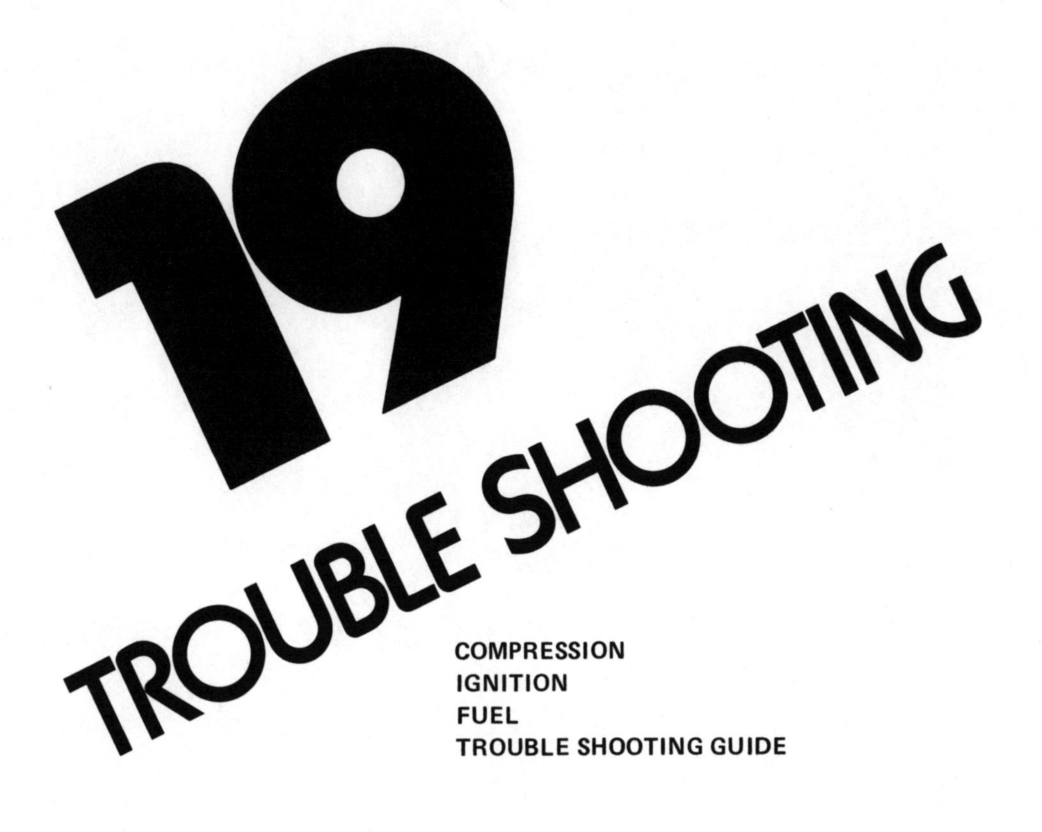

19 TROUBLE SHOOTING

COMPRESSION
IGNITION
FUEL
TROUBLE SHOOTING GUIDE

What do you do when trouble occurs? The engine does not start. It stops and does not restart. Do not overlook possible causes that seem too obvious to be considered: an empty fuel tank, a closed fuel valve, or a switch that is not "on." Always check the simplest causes first.

Also, always keep these requirements in mind when trying to locate the cause of a problem: To operate, an engine must have (1) the correct air-fuel mixture, (2) a good spark to jump the spark plug gap, and (3) good compression to compress the air-fuel mixture.

For more complicated trouble, of course, detailed procedures for various checks and corrections on particular components are presented throughout the book. But a quick check procedure, which follows, enables you to readily eliminate components that are not causing the problem and thereby to isolate the unit that is causing trouble.

COMPRESSION

Compression is the pressure developed by the air-fuel mixture within the combustion chamber. You can check compression by pulling on the starting rope slowly until you feel the greatest resistance. Slack off on the rope and note the amount of snap-back. You can also pull and slacken the starter rope to determine the "bounce" off the high end of the compression build-up. If there is little or no bounce or snap-back, the trouble can be traced to poor compression. If a starter rope is not used, try turning the engine over by hand and note the resistance or "bounce." Turn the engine in the direction in which it operates. If a compression release valve is used, the engine must be turned in the opposite direction to determine compression. To get an accurate reading use a compression gauge.

IGNITION

To check the ignition system remove the spark plug from the cylinder head and then reconnect the spark plug cable to the spark plug. Hold the threaded part of the spark plug base against clean bare metal on the engine. With the ignition switch "on," crank the engine briskly. If there is no spark between the spark plug electrodes, replace the spark plug with one you know is good. Turn the engine over; if no spark is present, the ignition system is at fault. If a battery ignition system is used, the battery must be charged to some degree and the electrical connections must be good to make a spark.

FUEL

Most carburetor operating troubles are caused by the improper adjustment of the main fuel or idle mixture adjustments.

To check for fuel reaching the combustion chamber, remove the spark plug and close the choke valve. Crank the engine over several times while holding your thumb tightly over the spark plug hole. If your thumb does not become moist with fuel, then no fuel is reaching the combustion chamber. Be sure the spark plug cable is placed out of the way while cranking the engine. Touching the cable terminal while turning the engine with the ignition on will give you a shock.

Check to determine if fuel is reaching the carburetor. If a gravity feed system is used, fuel should run out of the line when it is disconnected at the carburetor—assuming the shut-off valve is open. If a separate pump is used, crank the engine. Fuel should flow out of the line from the pump.

TROUBLE SHOOTING GUIDE

The following list of problems and their causes applies in general to both four-stroke and two-stroke cycle engines. Hence the phrase "Wrong fuel mixture" refers to the mixture of gasoline and oil in a two-stroke cycle engine, but "Improper air-fuel mixture" refers to the air-fuel ratio controlled by the idle and main fuel adjustment.

In most cases a number of conditions could be the cause of the problem. So always check and correct the simplest problem first.

Problem	Causes
Engine turns over but does not start	No fuel No spark No compression
Engine does not turn	Seized engine—lack of lubrication Hydrostatic lock (two-stroke cycle)—fuel has entered crankcase, flooding engine Broken internal parts Dead battery, if electrically operated starter
Engine skips or misfires	Wrong fuel mixture Poor ignition Spark plug fouled or gap too wide Ignition timing Improper air-fuel mixture
Engine lacks power	Engine overload Improper air-fuel mixture Wrong fuel mixture Weak spark Fuel filter partially clogged Carbon build-up (two-stroke cycle) Poor compression Too hot Muffler clogged Bind or drag in equipment engine operates
Engine backfires	Improper air-fuel mixture (lean) Valve problem Ignition breaker point gap Improper timing setting

Problem	Possible cause
Engine starts hard, stalls on idle, operates well at high speed (two-stroke cycle)	Air leak into crankcase Cracked crankcase Gaskets not sealing Reed valve broken Fittings on crankcase leaking Oil seals leaking
Engine hard to start	Improper air-fuel mixture Wrong fuel mixture Faulty fuel pump Choke valve not operating properly Weak ignition (point setting) Head gasket, valves, or rings leaking—poor compression Drag caused by equipment Engine too hot Spark plug gap too wide Ignition wires loose or shorting Weak ignition Plugged fuel line or filter
Engine stops suddenly	No fuel No spark No lubrication (engine seized) Valve problem
Engine operates erratically (surges)	Wrong fuel mixture Improper idle adjustment Engine overload Governor faulty Clogged carburetor passages Vent cap on fuel tank plugged Water in fuel Weak ignition (point gap) Clogged fuel line and/or filter
Engine idles poorly	Wrong fuel mixture Improper idle adjustment Poor ignition (point gap) Air leak into intake system Leaking head gasket
Engine knocks or pings	Gasoline octane rating too low Improper air-fuel mixture Engine over loaded Carbon build-up Advanced ignition timing Engine too hot Wrong fuel mixture

Engine overheats	Wrong fuel mixture Retarded ignition timing Improper cooling Engine overload Air inlet opening clogged Oil level too low or too high Muffler obstructed Improper air-fuel mixture (lean) Blocked cooling fins
Engine knock	Lack of lubrication Engine too hot Loose crankshaft bearings Excessive valve clearance Loose flywheel Excessive carbon build-up
Knock in drive assembly	Chipped gear teeth Broken gear teeth Loose shafts
Whine or noise in driven units	Worn bearings or bushings Excessive gear clearance Loose pulley Loose blade on mower
Vibration	Loose mounting bolts Mower—loose or out-of-balance blade Frayed belts Loose pulley

20

CHARACTERISTICS OF SPECIFIC UNITS POWERED BY SMALL ENGINES

Generally different types of machines come in great enough varieties that you can find the equipment to best fit your requirements. By becoming familiar with available options, you will be able to make a better selection. This chapter includes a general discussion of the more popular types of machines that are powered by small gasoline engines. It also gives you information about their characteristics, such as the operating information and features to consider when comparing different brands. Some operating safety features are also discussed.

CHAIN SAWS

Chain saws are available in a variety of sizes. Guide bar lengths are available from 10 inches to over 27 inches, but the larger sizes are generally used by commercial loggers and tree trimmers. The most widely used saws are smaller, and individuals use them to cut up wood and to remove tree limbs. Regardless of their sizes, however, the basic construction of all chain saws is very much alike.

The weight of the unit and the length of the guide bar are two factors to consider when purchasing a saw. Generally the length of the guide bar, which supports the chain, is the factor that determines the horsepower (size) of the engine. The general—and obvious—rule is: the longer the guide bar, the greater the diameter of log that the saw can cut. But because you can cut around the log, you can "double" the length of the guide bar; a 12-inch guide bar permits you to cut a 24-inch log, 12 inches from either side. Figure 20-1 is an illustration of a typical chain saw.

Gasoline chain saws use a two-stroke cycle engine with magneto ignition. A rewind (recoil) rope starter is used to start the engine. A centrifugal clutch engages the cutter chain when the engine reaches a predetermined speed, approximately 3,000 rpm.

Since the cutter chain must be kept lubricated at all times, chain saws are generally equipped with an automatic oiler. A small container (tank) is built into the unit for a special chain oil, and an oil pump delivers a supply of oil to the chain when the saw is operating. The oil reduces friction and wear from between the guide bar and chain. There is usually an adjustment to control the oil flow.

As the saw is used, the chain stretches, as well as wears to some extent. To compensate for this stretch, the guide bar can be adjusted to take up any slack between the chain and guide bar. The cutting teeth on the chain become dull with use. To function properly, the saw's cutting teeth must be kept sharp. Correctly sharpened, the chain cutting teeth slice wood easily with a low feeding pressure.

The cutting teeth on the chain have a definite sharpening angle, usually 30° to 35°, which must be maintained for effective cutting. The teeth can easily be sharpened, but only with a special chain-sharpening file. Keep *all* the cutter teeth at the same length. When sharpening, find the shortest cutter tooth first and use this as a reference for sharpening the others. Sharpen all the cutters on one side, and then sharpen the other side. The simplest way to correctly file a chain saw cutter is to use a file holder or a filing tool with markings that show the sharpening angles. Figure 20-2 shows sharpening the saw cutting edge with a file holder.

FIGURE 20-1
Chain saw with parts labeled
Courtesy Stihl Corporation

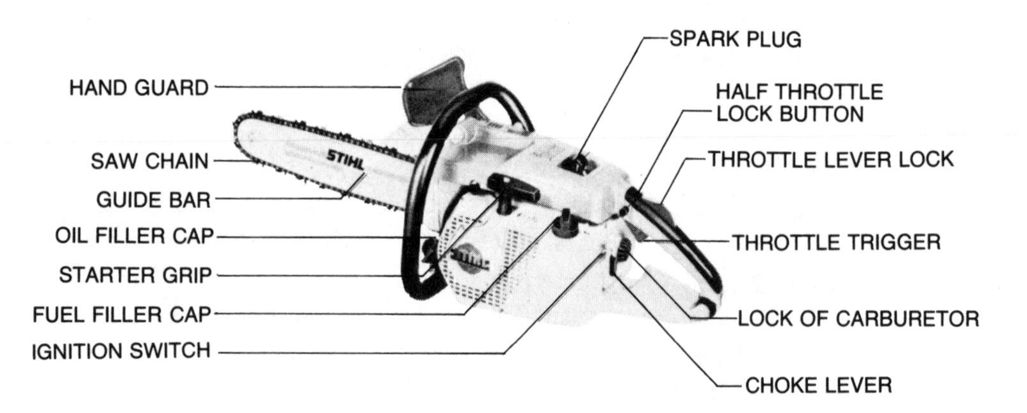

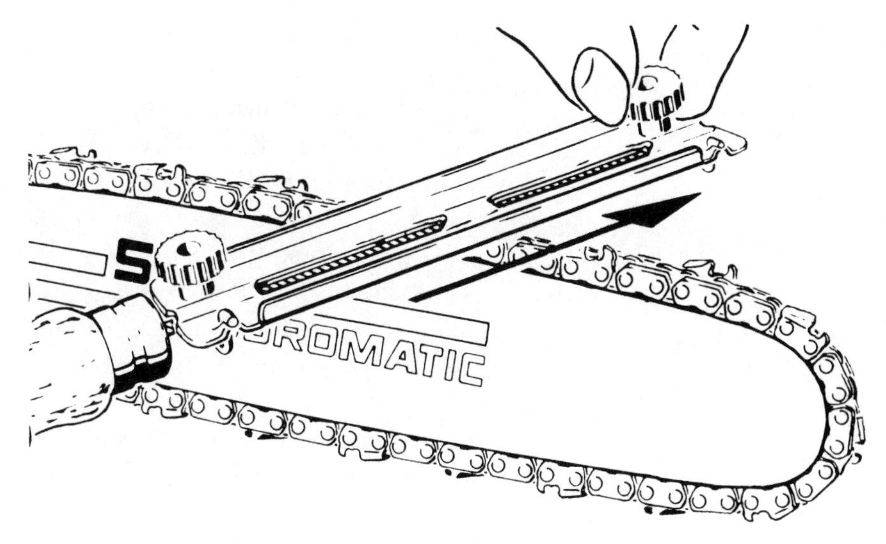

FIGURE 20-2
Sharpening with a file holder
Courtesy Stihl Corporation

A chain brake is standard on many saws. Its function is to apply pressure to the clutch drum, which disengages the clutch from the chain, thus bringing the chain to a rapid stop when the brake is applied. The brake is triggered when the saw kicks back and the operator's hand hits the hand guard.

When operating the saw, always keep the chain clean and regularly remove the sawdust and chips from the engine housing openings.

A slightly different starting procedure is necessary for most chain saws. With an automatic chain brake, make sure the brake is reset. Move the ignition stop switch away from "stop." Place the saw on firm ground or some other solid surface in an open area. Be sure that the chain is clear of all obstructions and objects, because the clutch engages the chain when the engine starts.

A cold engine is started with the choke applied (the valve closed). Squeeze the throttle trigger and push the throttle lock in, thus locking the throttle at about the half-open position. Hold the saw down with your hand and place your foot through the handle. Pull the starter rope slowly until you feel the starter engage. Give the rope a quick pull. Do not let the rope snap back. Crank the engine until it fires. Push the choke in so the choke valve is open. Crank the engine until it starts. As soon as the engine starts, depress the throttle trigger to release the throttle stop. Let the engine idle.

When the engine is warm, it should start without using the choke.

LAWN MOWERS

Lawn mowers, in their many sizes and types, are generally divided into reel and rotary types and then classified according to the width of their cuts, as well as according to whether they are for riding, self-propelled, or hand-pushed. The major difference in mower construction is between the reel and rotary types.

Although the *reel mower* does a very good mowing job because it shears the grass, it does not cut through tall grass very well, and it works best on even ground. The cutting blade must be kept sharp and carefully adjusted for satisfactory operation. The reel mower is also more expensive because it includes more operating parts.

The blade of the *rotary mower* creates a suction, which stands the grass up so it is cut in a machete fashion by a single high-speed spinning blade. Except on large mowers, where more than one blade is used, the blade is attached directly to the engine crankshaft. The rotary mower can cut high grass and does not demand an even terrain. It has fewer parts than the reel type mower.

The width of the mower cut may range from 18 to 34 inches for a rotary mower. Larger mowers are available, but they are usually "add-ons" to a lawn or garden tractor. The most common size mower is in the 18- to 22-inch range. Riding mowers have a cutting width in the 24- to 34-inch range.

Although a few mowers are equipped with two-stroke cycle engines, most use four-stroke cycle engines. Their horsepowers range from about 3 to as high as 10 for a riding mower. Most riding mowers have at least a 5-horsepower engine.

Cutting heights are adjustable and may range from approximately 1 to 4 inches. Raising and lowering the wheels is generally the method used to adjust cutting height. The adjustment lever may be located on the control handle or on the mower deck. On some mowers the cutting height is changed by unbolting the wheels and putting the attaching bolt in a higher or lower hole.

Self-propelled mowers may use either the front or the rear wheels to propel the mower. The power from the engine may be transferred to the wheels by means of a belt or chain. The power may be engaged or disengaged by tightening or loosening the belt. A friction clutch may be used to apply power to the wheels through a belt or chain.

Grass clippings left on the lawn do not look good and, if heavy enough, may harm the grass. Most rotary mowers are constructed so a grass catcher can be attached to the mower housing. Some use a side bagger, which cuts down on maneuverability in tight places, while others use a rear bagger. The rear bagger may result in dust filtering back toward the operator. A mulching attachment, available for some mowers, cuts the clippings extra fine and deposits them evenly on the lawn as a mulch.

Although riding mowers are designed strictly for cutting grass and are not intended to operate other equipment, some people use them to pull lawn carts.

With so many mower options available, assess the job that needs to be done before deciding on which mower to purchase.

One of the most important considerations in the size of the

lawn. If you have over 5,000 square feet of lawn, or if your property is hilly, consider a self-propelled mower. Naturally the least expensive mower is the push-type with a narrow cutting width and no grass-catching attachment. When considering the push-type, remember that the lighter ones are easier to push!

In addition to the size of the lawn, other factors are important: Consider the trees, shrubs, and flower beds that you must mow around. Is the terrain smooth or hilly? How frequently does the lawn need to be cut? What is the type of grass? And how short should it be cut? Also, are you physically able to push a mower? You need to consider whether you want to collect the clippings, mulch them, or leave them on the lawn.

Numerous options are available from different manufacturers. You may want to consider some of these factors when purchasing a mower:

Push-type, riding, or self-propelled
Electric start or recoil rope start
Bagger, side or rear, no catcher, mulcher
Front- or rear-wheel drive
Cutting width
Cutting height adjustment on handle, on platform, or by bolts
Two-stroke or four-stroke cycle engine
Horsepower range

The rotary mower is equipped with a removable blade, which can be replaced or sharpened. If not kept balanced, the blade may cause vibration. Be careful when sharpening the blade so as to retain the balance. Do not remove a lot of material from just one half of the blade.

Never turn the blade by hand unless the spark plug wire is disconnected. The engine may start or kick back, causing injury to you.

Always keep the mowing section clean and free from grass. After mowing, tip the mower up and spray the underside of the mowing section with the garden hose to wash the clinging grass and dirt from the housing.

Figure 20-3 is a cut-away view of a hand pushed mower.

Safety Precautions For Mowers

1. Keep your hands and feet away from the cutting blade when starting as well as when operating.
2. Keep everyone away from the grass discharge. The moving blade can pick up a stone and shoot it out of the grass discharge opening literally like a bullet.

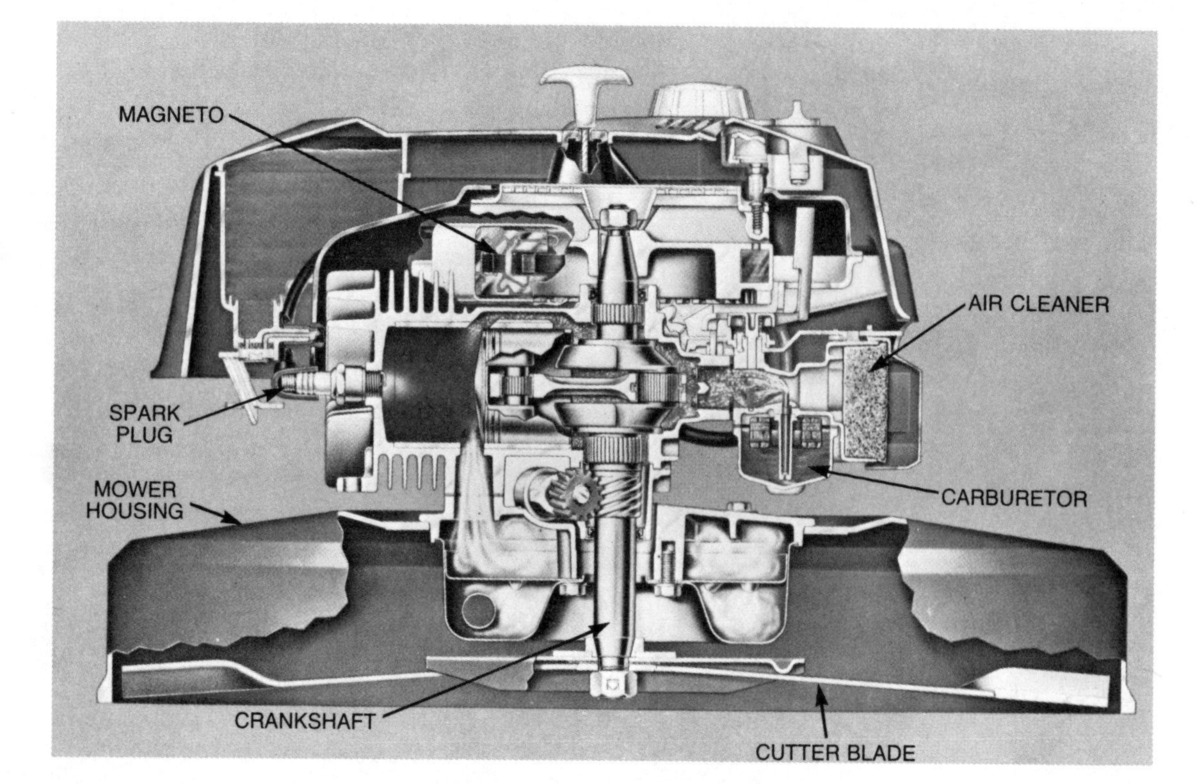

FIGURE 20-3
Cut-a-way view of hand-pushed mower
Courtesy Gale Products, Division of Outboard Marine Corp.

3. Special attention must be paid to yard hazards, such as steep banks, ditches, holes, low-hanging branches, and other obstructions. Mow steep slopes up and down not across.
4. Do not refuel the mower when the engine is running or hot. The gasoline can explode.
5. When operating a riding mower, pay special attention to additional safety factors.
6. Do not run the machine unless sitting in the driver's seat.
7. Keep your hands and feet away from the blade housing.
8. Disengage the mower before starting the engine.
9. Do not let your feet touch the ground when the mower is moving.
10. Never try to unclog the discharge chute until the engine is shut off and the blade has stopped.
11. Stop the blade while crossing gravel driveways.
12. Never leave a running machine unattended.

LAWN AND GARDEN TRACTORS

Most lawn and garden tractors that are powered by single-cylinder gasoline engines are in the 6- to 16-horsepower range. These versatile machines are designed to pull and/or to operate various pieces of equipment. Their most common use is for the maintenance of lawns and home gardens. Most are equipped with a power take-off to operate different pieces of equipment such as a mower, snow blower, and the like.

Remember this when buying a tractor: The tractor by itself is of little use. It is designed to pull working equipment and to power attached equipment. So you need various pieces of equipment to perform the work you want done.

The most common piece of equipment is the rotary mower, which can usually be mounted beneath the tractor and which becomes part of the machine. The blades are normally driven by a power take-off pulley on the engine. The mower attachment can be removed when the tractor is to be put to other uses, or it can be disengaged so it does not operate when using the tractor.

Another common attachment, a scraper blade is generally attached to the front of the tractor. Although its primary use is to push snow off the driveway and sidewalk, it can also be used to push dirt.

Remember that these lawn and garden tractors cannot perform the heavy work that a farm tractor performs. With multi-ratio transmissions, many tractors have ample power, but they do not have enough weight to provide the necessary traction to move big snowdrifts, to push large piles of dirt, or to pull a large plow. If you must consider moving a lot of snow, obtain a snow blower attachment. The tractor propels the machine as well as operates the blower from the tractor power take-off.

Most tractors with 8-horsepower or more can handle a rotary mower, vacuum grass catcher, small snow plow, dozer blade, snow blower, dump cart, leaf sweeper, sprayer, rake, and fertilizer spreader. Ten or more horsepower is generally required to effectively handle a tiller, cultivator, harrow, small (10-inch) plow, rear scoop, or a grader blade.

There are many options to consider when purchasing a lawn or garden tractor. Most of the lower-priced tractors are equipped with 3-speed manual transmissions, while others have a 4-, 5-, or even 6-speed manual transmissions. Some manufacturers use a hydrostatic (hydraulic) transmission, which has infinitely variable ratios and requires no shifting other than forward and reverse. Other tractors may be equipped with a semi-automatic drive, which reduces the amount of shifting required.

A limited slip differential is available on some tractors. Instead of the wheel that has traction standing still and the wheel that is not gripping spinning, the wheel with traction will turn. The free wheel stands still, they both turn as the load becomes equal.

Some tractors have power take-offs located at both the front and rear of the machine. Others have the unit near the center of the tractor. In some cases the power take-off pulley operates whenever the engine is running; others have a control lever so the unit can be engaged or disengaged independently of the engine.

An auxillary transmission may be avilable if the tractor is to be put to heavier-than-normal use. In some cases different rear axle ratios are available for additional power or speed. Different tire and wheel sizes may also be available to tailor the machine to better fit your needs.

A hydraulic pump attachment, made available by some manufacturers, enables you to operate a hydraulic lift unit, to raise and lower equipment and to operate hydraulic equipment such as sprayers, front end loaders, and the like.

Most tractors are equipped with an electric starter, while others have a rope recoil mechanical starter. Some have a battery ignition system, and others use a magneto-type ignition system. Headlights are available for many tractors.

From the available options you can have a tractor tailored to fit any particular need. Shop around and compare a number of different makes of tractors, as well as the options available.

For ordinary lawn and garden care, an 8- to 10-horsepower tractor, with a mower attachment and a dozer blade for snow removal, serves most people. When buying a particular tractor, check to see what equipment is available for it. At some later date you may wish to get additional equipment. If the equipment is made for a particular tractor, it usually serves you better. Figure 20-4 shows a typical lawn and garden tractor.

FIGURE 20-4
Typical lawn and garden tractor
Courtesy Engineering Products Co.

Transmission may be of somewhat different designs, but they all perform in much the same manner. The frames and suspension systems may be heavier on some machines than on others, but added weight means poorer gasoline mileage.

A variable sheave drive pulley, standard equipment on some machines, gives a smooth flow of power and a variable drive ratio but an increase in the cost of the machine. Different-sized tires may make for a more comfortable ride, as do air cushion suspension, which is available from some manufacturers.

When buying any machine or piece of equipment, always look over as many different makes and models as possible. Take a demonstration ride. Compare the features you like so you can determine which best fits your need.

SNOWMOBILES

Snowmobiling has become a very popular means of recreation in some sections of the country. To take full advantage of the machine, you need ample snow most of the winter and suitable terrain. "Suitable" means open spaces without fences, or at least trails that have been established for the use of snowmobiles. As time goes on, more areas are being opened up for the operation of snowmobiles. Figure 20-7 shows a typical snowmobile.

While this book is limited to single-cylinder engines, all of which are air-cooled, the high-performance large snowmobiles use 2- or 3-cylinder engines, some of which are liquid-cooled. Yet a number of snowmobiles in use employ a single-cylinder air-cooled engine.

FIGURE 20-7
Typical snowmobile
Courtesy Polaris E-Z-Go Division of Textron Inc.

The engines, regardless of the number of cylinders, are of the two-stroke cycle type. Most, except for the larger engines, are equipped with a recoil-type starter, although electric starters may be available as optional equipment for some engines. Engine sizes may vary from approximately 250 cc displacement for the single cylinder engines to 500 cc for the multi-cylinder engines. The horsepower ranges from 20 up to about 50. The weights of the different machines range from about 275 pounds up to 400 pounds, depending on the horsepower and accessories.

Most snowmobiles are equipped with hydraulic disc brakes, with the controls usually mounted on the handlebars. The track, made of molded rubber, may or may not be reinforced with fiberglass. Some tracks have rubber cleats, while others use metal cleats. Usually the cleats can be replaced when they become worn.

The drive system on most snowmobiles is similar to that found on many mopeds. The snowmobile uses the drive and driven pulleys (clutch) as a torque multiplier. This arrangement also serves to engage and disengage the track from the engine. (Some mopeds do not use the clutch as a torque converter.)

On this type of system the drive pulley is attached to the engine crankshaft. The drive pulley (clutch) cover is moved in and out by spring tension and centrifugal weights according to engine speed. As the cover, which is one side of the pulley, moves in, the belt rides higher in the pulley, causing the belt to turn faster. The driven pulley (clutch) is attached to the drive shaft. The sprockets that drive the track are mounted on the drive shaft. The driven pulley has a movable sheave (side) that is held close to the stationary side of the pulley by spring tension at low speeds. At high belt speed, the sheave is forced apart so the belt rides lower in the pulley, causing the shaft to turn faster. With this arrangement no gears are needed to provide power to meet the varying operating conditions.

Certain precautions should be observed when working around a snowmobile, as well as when operating the machine.

1. When checking engine operation, do not run the engine with the clutch guard removed. Should the clutch fail, flying metal parts could cause you serious injury.

2. When warming up the engine or clearing the track of snow, make sure no one is standing in front or behind the machine.

3. Do not operate the engine with the carburetor intake silencer and cleaner removed. Doing so can result in damage to the engine.

4. The machine is designed to operate best on unpacked snow. Maneuverability is attained by the steering skis and by shifting your body weight. Maneuverability is lessened under packed snow or icy conditions, so it is dangerous to operate on glare ice.

5. Zig-zag up or down steep hills. Always cut back on the throttle to gear down for more power going up hills.

6. Do not operate for prolonged periods on black top, gravel, or glare ice. Doing so causes excess wear on the track.

When purchasing a new machine, always compare the various available machines. Test drive the different machines and decide which best fits your needs for the price you can pay.

GOLF CARTS

Golf carts are becoming more prevalent as additional people play golf. Besides their popularity on most golf courses, a number of retirement communities permit their use within the confines of certain areas. They are an inexpensive means of off-highway transportation. Factories, shopping malls, and other large areas use them for transporting personnel as well as small packages. Because of the exhaust fumes gasoline carts should not be used inside areas where people congregate.

Golf carts may be of the electric or gasoline type. The electric cart depends on storage batteries, which must be charged from an outside electrical source, for power. The gasoline cart is self-contained, economical, and easy to operate. The carts are restricted to two occupants and their equipment. Our discussion is confined to the gasoline cart.

The gasoline cart is powered by small one-cylinder engines, most of which have approximately 20 to 30 cc displacement. The cart is simple to operate. To drive, turn the switch key on and depress the accelerator. The accelerator pedal starts the engine automatically, and further movement operates the cart at the desired speed. A choke, operated by a pull button on the panel, may have to be pulled to start a cold engine and moved inward as the engine warms up. The engine stops when the accelerator pedal is released.

A pedal-operated disc or drum brake is used to stop the cart. Tilting the pedal pad forward holds the brake in the applied position for parking purposes. The brake releases automatically when the accelerator is depressed. Some carts use a hand-operated brake lever to apply and to hold the brake for parking purposes.

A forward and reverse lever is used to select the direction of travel.

The cart is powered by a two-stroke cycle air-cooled engine. Top speed is generally limited to 15 mph. Reverse is obtained in many carts by reversing the direction of engine rotation. Some carts obtain reverse through the use of a sliding drive gear assembly in the rear axle.

An automatic transmission, used on most installations, provides for neutral and a variable ratio. The machine is not designed for the engine to run in neutral, so neutral is used to shift from forward to reverse. A rear wheel must be jacked up in order to run the engine without moving the cart. The variable ratio is obtained through a variable pitch sheave pulley arrangement. Ratio changes take place as the diameter of the drive surfaces on the pulley becomes larger or smaller according to load and speed.

A motor/generator is used to crank the engine to start and to charge the battery. All the newer carts use a 12-volt battery for the ignition system and starter.

The rear axle assembly is usually like the automotive type. Very little service is necessary.

The carts may be available with three- or four-wheel suspension. The four-wheel cart has greater stability than the three-wheel type, some of which use a tiller for steering. Most steering gear mechanisms are of the automotive type, as is the steering linkage on the four-wheel carts.

Coil or leaf springs are used in the suspension system. Automotive type shock absorbers are used to complement the springs.

The following suggestions are relative to the operation and use of a golf cart:

1. Drive slowly when making sharp turns. It is easy to fall out if you are not careful.

2. Keep your arms and legs inside the cart when moving. Be sure the cart is not moving when you get out.

3. If possible, drive straight up and down steep hills.

FIGURE 20-8
Typical gasoline-powered golf cart
Courtesy Harley-Davidson Motor Co., Inc. Subsidiary of AMF, Inc.

4. Try to park on level ground and set the parking brake. Do not use the accelerator to hold the cart; use the brake.

5. Do not leave the cart unattended unless the switch is turned off. With the switch on, merely depressing the accelerator causes the cart to move.

Figure 20-8 shows a typical gasoline-operated golf cart.

INDEX

(Index consolidated into one page for your easy reference.)